Disturbance Observer-Based Control

Methods and Applications

Disturbance Observer-Based Control

Methods and Applications

Shihua Li • Jun Yang
Wen-Hua Chen • Xisong Chen

CRC Press
Taylor & Francis Group
Boca Raton London New York

CRC Press is an imprint of the
Taylor & Francis Group, an **informa** business

CRC Press
Taylor & Francis Group
6000 Broken Sound Parkway NW, Suite 300
Boca Raton, FL 33487-2742

First issued in paperback 2017

ISBN-13: 978-1-4665-1579-6 (hbk)
ISBN-13: 978-1-138-19998-9 (pbk)

Library of Congress Cataloging-in-Publication Data

Li, Shihua (Engineer)
    Disturbance observer-based control : methods and applications / authors, Shihua Li, Jun Yang, Wen-Hua Chen, Xisong Chen.
        pages cm
    Includes bibliographical references and index.
    ISBN 978-1-4665-1579-6 (hardback)
    1. Adaptive control systems. 2. Observers (Control theory) 3. Nonlinear control theory. I. Title.

TJ217.L48 2014
629.8'36--dc23   2013045985

Visit the Taylor & Francis Web site at
http://www.taylorandfrancis.com

and the CRC Press Web site at
http://www.crcpress.com

To my wife Xiamei Huang with appreciation and love. Without her durative support behind me, there is no way for me to maintain a stable tracking control toward the desired research target. To my daughter Huangyu (Lisa) Li with love. She always brings me happiness and inspiration.

Shihua (Sandy) Li

To my parents Heyu Yang and Aiqing Li, my wife Yanan Wei, and my son Haoyu Yang for their endless support and love.

Jun Yang

To my parents.

Wen-Hua Chen

To my wife and son.

Xisong Chen

Contents

SECTION I   OVERVIEW

SECTION II DISTURBANCE ESTIMATION DESIGN

SECTION III DISTURBANCE OBSERVER-BASED CONTROL DESIGN

List of Figures

List of Tables

Preface

As an urgent desire has come up for high-precision control in modern engineering systems, disturbance and uncertainty attenuation has become increasingly important and has attracted much attention in both academic and industrial communities. Traditional feedback-control schemes such as proportional-integral-derivative (PID) controllers, attenuate disturbances by feedback regulation in a relatively slow way. In the control community, it is well-known that feedforward control provides an effective disturbance compensation method that can achieve prompt disturbance attenuation. However, the implementation of traditional feedforward control requires that the disturbance has to be measured by sensors. Unfortunately, in many cases, the disturbances are difficult or even impossible to be measured physically by sensors. This is the main factor restricting the development and applications of feedforward control in control engineering.

To this end, developing disturbance estimation techniques would be a good choice to alleviate the restriction faced by traditional feedforward control. Disturbance observers (DO) provide an effective disturbance estimation technique for a wide range of systems. Disturbance observer-based control (DOBC) has been regarded as one of the most promising disturbance-attenuation approaches. In the authors' opinion, DOBC has the following three distinctive features: Firstly, disturbance observer-based compensation can be considered as a "patch" to existing controllers that may provide good stability and tracking performance but usually show unsatisfactory disturbance attenuation and robustness against uncertainties. The benefit of this is that there is no change to a baseline feedback controller which may have been widely used and developed for many years, for example, model predictive controller in typical process control systems and gain-scheduling controller in classical flight control systems. After a baseline controller is designed by using the existing feedback control methods, the disturbance observer-based compensation is added to improve the robustness and disturbance attenuation of the baseline controller. Instead of employing a completely new and different control strategy that demands a new verification and certification process, the verification of DOBC can be developed based on the existing verification process to ensure safety and reliability. Secondly, DOBC is not a worst-case-based design. Most of the existing robust control methods are worst case based design, and have been criticized as being "over conservative." In general, in those robust control methods, promising robustness is achieved at the price of degraded nominal performance. In the DOBC approach, the nominal performance of a baseline controller is recovered in the absence of disturbances or uncertainties. Finally, most disturbance-rejection approaches, such as PID control and robust control, suppress disturbances through feedback regulation, which may lead to a slow response in attenuating the influence of disturbance. Instead, by introducing a disturbance- compensation part, DOBC provides an active disturbance-rejection approach and is able to attenuate the disturbances promptly.

Although DOBC has been known as one of the most promising and effective methods by many active researchers and practitioners in the control community, so far, there are no academic books about DOBC. This book aims to introduce the DOBC technique to a wider range of audience, and promote its applications and research by explaining the basic principles and presenting new developments of DOBC in both theoretic and practical aspects. During the past 30 years, DOBC has achieved numerous successful applications in engineering due to its powerful abilities in compensating disturbances and uncertainties. The authors have engaged in the field of DOBC since 1999. In this book, we would like to present and summarize the research results in the field of DOBC in both theoretic contributions and practical engineering applications. In the theoretic aspect, this book will introduce a number of disturbance estimation and compensation methods for different type of systems with different type of disturbances, such as single-input–single-output (SISO)/multi-input–multi-output (MIMO) linear/nonlinear systems, time-delayed systems, and nonlinear systems with constant, harmonic, high-order, matched and mismatched disturbances. In the application aspect, the authors have contributed to the development of DOBC for practical control applications for many years (e.g., motion control, flight control, and process control). For example, Shihua Li's research group successfully applied the DOBC method to the product of AC servo motor systems in ESTUN AUTOMATION company in China, which provides a significant economic benefit for the company. Wen-Hua Chen's laboratory has successfully employed DOBC for the control design of Small-scale autonomous Helicopters. Xisong Chen's group has developed DOBC techniques for various process control systems, such as ball Mill grinding circuits and level tank systems.

The major topics of this book are about theoretic methods and practical applications of disturbance estimation and compensation for control systems through a DOBC approach. The book is arranged as the following parts: Section I provides an overview on disturbance observer-based control approach, including the history and features in both disturbance estimation and compensation aspects. Disturbance estimation is an essential part of DOBC. Previous disturbance estimation approaches can only handle very limited types of disturbances. A number of new disturbance estimation techniques will be introduced that are applicable for a wide range of disturbances and uncertainties. New developments on disturbance estimation techniques are presented in Section II. Traditionally, DOBC are only applicable for matched disturbance attenuation. In Section III, both mismatched and matched disturbance/uncertainty attenuation for DOBC will be presented. The disturbance observers are integrated with a baseline controller designed with existing well established control methods such as feedback linearization control and sliding model control. In Sections IV-VI, applications of the methods provided in Sections II and III are illustrated by presenting various practical engineering case studies, which would be particularly attractive for the control engineering practitioners who are interested in disturbance attenuation and uncertainty compensation.

The minimal skill set to adequately read and follow the book is the undergraduate level of basic automatic control theory. The readers with the knowledge of basic control theory shall be able to understand most parts of the book, including the introduction of the DOBC concept and its applications. However, to understand advanced materials presented in this book, in particular, nonlinear control methods introduced in Sections II and III, it would be beneficial for the reader to have preliminary knowledge about nonlinear control. We believe that this book will provide guidance for both engineers who are interested in disturbance estimation and compensation techniques for practical engineering systems, and research scholars who are interested in theoretic research and methodology development in related fields.

Acknowledgments

The main contents of this book are the results of extensive research efforts in the last 10 years. Some of the results have been reported in the research papers [1, 2, 3, 4, 5, 6, 7, 8, 9, 10, 11, 12, 13, 14, 15, 16, 17, 18]. We would like to express our thank to our collaborators Peter J. Gawthrop, Donald J. Ballance, John O'Reilly, Lei Guo, Xinghuo Yu, Argyrios Zolotas, Qi Li, Cunjia Liu, and Konstantinos Michail for their active discussions and valuable suggestions when we carry out research in this area. We are grateful to our colleagues, in particular, Zhiqiang Gao, Yi Huang, Xin Xin, Xinjiang Wei, Mou Chen for their enthusiastic helps during the research process. Our thanks also go to the former and current students Huixian Liu, Zhigang Liu, Shihong Ding, Haibin Sun, Juan Li, Zhengxin Zhang, Kai Zong, Xiaohua Zhang, Cheng Chen, Chao Wang, Xiaocui Wang, Xuan Zhou, Hao Gu, Mingming Zhou, Jinya Su, Hongchao Wang, Cong Guo, Cunjian Xia, Shuna Pan, and Jiejin Yan from School of Automation at Southeast University for their diligent and active works on applications of DOBCs to various industrial systems. The second author would like to thank the China Scholarship Council (CSC) for supporting his visit to UK. He also acknowledges the Department of Aeronautical and Automotive Engineering for hosting his research at Loughborough University.

Most parts of this work would not have been possible without the financial support from several sources. The first author expresses thanks to the support of the National 863 Project of the Twelfth Five-Year Plan of China under Grant 2011AA04A106 and New Century Excellent Talents in University under Grant NCET-10-0328. The second author's work was supported by National Natural

Science Foundation of China under Grant 61203011, and Natural Science Foundation of Jiangsu Province under Grant BK2012327. The third author also would like to express his thanks to UK Engineering and Physics Science Research Council under Grant EPSRC GR/N31580.

Nanjing, Loughborough

Shihua Li
Jun Yang
Wen-Hua Chen
Xisong Chen

The Authors

Shihua Li was born in Pingxiang, Jiangxi Province, China, in 1975. He received his B.S., M.S., and Ph.D. degrees all in automatic control from Southeast University, Nanjing, China, in 1995, 1998 and 2001, respectively. Since 2001, he has been with the School of Automation, Southeast University, where he is currently a professor. His main research interests include nonlinear control theory with applications to robots, spacecraft, AC motors, and other mechanical systems.

Shihua Li, PhD
School of Automation, Southeast University, No.2 Si Pai Lou,
Nanjing 210096, China

Jun Yang was born in Anlu, Hubei Province, China, in 1984. He received his B.S. degree in the Department of Automatic Control from Northeastern University, Shenyang, China, in 2006. In 2011, he received the Ph.D. degree in control theory and control engineering from the School of Automation, Southeast University, Nanjing, China, where he is currently a lecturer. His research interests include disturbance estimation and compensation, advanced control theory with applications to flight control systems, motion control systems, and process control systems.

Jun Yang, PhD
School of Automation, Southeast University, No.2 Si Pai Lou,
Nanjing 210096, China

Wen-Hua Chen received his M.S. and Ph.D. degrees from the Department of Automatic Control at Northeast University, China, in 1989 and 1991, respectively. From 1991 to 1996, he was a lecturer and associated professor in the Department of Automatic Control at Nanjing University of Aeronautics and Astronautics, China. He held a research position and then a lectureship in control engineering at the Centre for Systems and Control at the University of Glasgow, UK, from 1997 to 2000. Currently, he is a professor in autonomous vehicle, and the head of the Control and Reliability Group in the Department of Aeronautical and Automotive Engineering at Loughborough University, UK. He has published more than 150 articles in journals and conferences. His research interests include the development of advanced control strategies and their applications in aerospace engineering.

Wen-Hua Chen, PhD
Department of Aeronautical and Automotive Engineering,
Loughborough University, Loughborough, LE11 3TU, UK

Xisong Chen received his B.S. degree from the School of Electronic Science and Engineering at Jilin University, China, in 1992. He received his M.S. and Ph.D. degrees from the School of Automation at Southeast University, in 2002 and 2009, respectively. Since 2002, he has been with the School of Automation, Southeast University, where he is currently a professor. His research interests include advanced control strategies and disturbance attenuation methods and their applications in process control engineering.

Xisong Chen, PhD
School of Automation, Southeast University, No.2 Si Pai Lou,
Nanjing 210096, China

OVERVIEW

I

Chapter 1

An Overview

1.1 Introduction

Disturbances widely exist in modern industrial control systems and bring adverse effects to the performance of control systems. Therefore, disturbance rejection is one of the key objectives in controller design. In this book, the disturbances refer to not only the disturbances from the external environment of a control system but also uncertainties from the controlled plant including unmodeled dynamics, parameter perturbations, and nonlinear couplings of multivariable systems, which are difficult to handle [19].

In the process control community, especially in petroleum, chemical, and metallurgical industries, production processes are generally influenced by external disturbances such as variations of raw material quality, fluctuations of production load, and variations of complicated production environment. In addition, the interactions between different production processes are always sophisticated and difficult to analyze precisely. These factors and their composite actions usually result in significant degradation of the production quality of these processes [20]. Taking the grinding and classification process of concentrator plant in the metallurgical industry as an example, the fluctuations of the feed ore particle size and the ore hardness are the external disturbances, which always cause the continuous fluctuations of the product particle size and further affect the production quality [21, 22, 1].

In the mechanical control community, including industrial robotic manipulators [2, 23, 24, 25, 26], motion servo systems [27, 28, 29, 30, 31], maglev suspension systems [32, 33, 34, 35, 36, 37], and disk drive systems [38, 39], etc., the control precision is generally affected by different external disturbances, such as uncertain torque disturbances, variations of load torque, vibrations of horizontal position of rail track, and pivot frictions. Moreover, the control performances of these mechanical systems are also subject to the effects of internal model parameter perturbations caused by the changes of operation conditions and external working environments [26, 28, 33, 39].

Also in the aeronautic and astronautic engineering community, applications such as missile systems [40, 41, 42, 43, 44, 45, 46, 47, 48, 49, 50, 51, 52, 53, 54, 55], near-space hypersonic vehicles [56, 57, 58, 59, 60, 61, 62], and space satellites [63, 64, 65, 66, 67, 68], the existence of disturbance forces and torques caused by external winds and unmodeled dynamics will severely influence the control performance. Furthermore, the practical dynamics of flight control systems will inevitably experience severe parameter perturbations due to harsh and complex flight environment, which brings great challenges to modern flight control design [40, 42, 57, 62, 63].

As analyzed above, unknown disturbances and uncertainties widely exist in various engineering control systems. The existence of these uncertain factors has great side effects on practical engineering systems. The problem of disturbance rejection is an everlasting research topic since the appearance of control theory and applications. Since no direct disturbance rejection design is considered, the traditional control methods, such as proportional-integral-derivative (PID) and linear quadratic regulator (LQR) controllers, may be unable to meet the high-precision control specifications in the presence of severe disturbances and uncertainties [4]. The essential reason for this is that these traditional methods do not explicitly take into account disturbance or uncertainty attenuation performance when the controllers are designed.

Therefore, the development of advanced control algorithms with strong disturbance rejection property has great importance to improve the control precision and of course the production efficiency of practical engineering systems. Due to the significance of disturbance attenuation, many elegant advanced control approaches have been proposed to handle the undesirable effects caused by unknown disturbances and uncertainties since 1950s. The major features of the typical disturbance attenuation methods are summarized as follows:

Adaptive Control (AC): AC aims to handle the undesired effects caused by structure parameter perturbations. The idea of AC is that the model parameters of the controlled systems is first identified online, then the control parameters are tuned based on the identified model parameters to obtain fine performance. AC is quite effective in dealing with model parameter uncertainties and has gained wide applications in practical engineering. The successful applications of AC methods usually depend highly on the design of identification or estimation laws on time-varying model parameters. When these key parameters are difficult to identify or estimate online, these methods are not available.

Robust Control (RC): RC focuses on researching the abilities of the controller against model uncertainties, which is an important branch of modern control theory. The control design of RC conservatively considers the worst case of model uncertainties. The robustness of RC is generally obtained at the price of sacrificing the transient performance of other featured points. Therefore, RC is often criticized to be overconservative.

Sliding Model Control (SMC): SMC has fine abilities in suppressing the effects of parameter perturbations as well as external disturbances. However, the discontinuous switching of the controller is prone to induce high-frequency chattering of mechanical systems. Although the employment of some modification methods such as the saturation function method could effectively reduce the chattering problem, the advantage of prominent disturbance rejection performance is sacrificed. Such disadvantages severely constrain the applications of SMC.

Internal Model Control (IMC): Since the early 1980s, the principal of IMC proposed by Garcia and Morari has been used to attenuate the effects of external disturbances in control systems. IMC has received a great deal of attention in both the control theory and application fields due to its simple concept and intuitive design philosophy. However, the IMC is generally available for linear systems. In addition, the implementation of IMC algorithm for high-dimensional system is quite sophisticated due to the requirement of calculating the inverse of a high dimensional transfer function matrix.

The motivation of the above-mentioned control approaches is to reject disturbances by feedback control rather than feedforward compensation control. These control approaches generally achieve the goal of disturbance rejection via feedback regulation based on the tracking error between the measured outputs and their setpoints [4]. Thus, designed controllers can not react directly and fast enough in the presence of strong disturbances, although they can finally suppress the disturbances through feedback regulation in a relatively slow way [20, 4]. To this end, these control approaches are generally recognized as passive antidisturbance control (PADC) methods.

To overcome the limitations of PADC methods in handling the disturbances, people have proposed the so-called active antidisturbance control (AADC) approach. Generally speaking, *the idea behind the AADC is to directly counteract disturbances by feedforward compensation control design based on disturbance measurements or estimations.*

Traditional feedforward control (FC) is known as the earliest AADC method. In the framework of FC, a sensor is firstly employed to measure the disturbances; secondly, the model of disturbance channel is built; finally, a feedforward controller, which employs the disturbance measurement and both models of the process and disturbance channels, is designed to counteract the disturbances. FC is one of the most direct and active disturbance attenuation methods. However, the applications of FC is limited due to the following reason. In most of the practical industrial processes, the disturbances are unmeasurable, difficult to be measured, or measurable but the sensors are excessively expensive. For example, the feed ore hardness in the grinding and classification processes is unmeasurable, the temperature of molten steel in steel plant is hard to be measured by real sensors, and component analysis instruments in chemical processes are generally very expensive.

To take advantages of the FC in disturbance rejection and also overcome its disadvantages as stated above, the exploration of *disturbance estimation* techniques (also called *soft disturbance measurements*) has attracted a lot of interest. Disturbance observer (DO) is one of the most effective and popular disturbance estimation techniques. Disturbance observer-based control (DOBC) has received a great deal of attention in both the control theory and control engineering fields. The motivations of the research on DOBC approach will be discussed in the next section.

1.2 Motivations

We will start with a simple example to illustrate the major motivation of this book. Consider the following system:

$$\begin{cases} \dot{x} = -ax + u + d, \\ y = x, \end{cases} \tag{1.1}$$

where u the control input, x the state, y the interested controlled output, a the system parameter, and d the disturbance. Let y_r the desired value that the output is expected to achieve, which is usually called *setpoint* or *object value*. Without loss of generality, the setpoint value y_r is taken as a constant one in this example for simplicity, that is, $\dot{y}_r = 0$.

Defining the tracking error variable of the system as $e_y = y_r - y$, system (1.1) is equivalently depicted by

$$\begin{cases} \dot{e}_y = -ae_y - u - d + ay_r, \\ y = y_r - e_y. \end{cases} \tag{1.2}$$

> The control object here is to design a control law u in terms of the tracking error and the setpoint, that is, $u = u(e_y, y_r)$, such that the real output y achieves its desired setpoint y_r, that is, the tracking error e_y tends to zero as time goes to infinity.

The basic principals for several typical control methods to reject disturbances are explored in details as follows.

1.2.1 High-Gain Control

Proportional control is usually utilized to realize the control object, where the control law has been designed as

$$u = ke_y + ay_r. \tag{1.3}$$

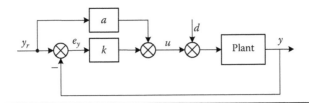

Figure 1.1   Block diagram of system (1.1) under the proportional control.

The block diagram of a closed-loop system (1.1) under the proportional control (1.2) is shown by Figure 1.1. Substituting control law (1.2) into system (1.2) yields

$$\dot{e}_y = -(a + k)e_y - d, \tag{1.4}$$

where k is the proportional gain to be designed.

To make the closed-loop system (1.4) stable, the control gain k in (1.2) has to be designed such that $a + k > 0$. It follows from (1.4) that

$$e_y(t) = e^{-(a+k)t}e_y(0) - \int_0^t e^{-(a+k)(t-\tau)}d(\tau)d\tau. \tag{1.5}$$

In the absence of disturbance, it can be concluded from (1.5) that $e_y(\infty) = 0$, which implies that the output will finally track its setpoint as time goes to infinity. However, in the presence of disturbance, we can not achieve similar results. Suppose that the disturbance in the system is bounded and satisfies $|d(t)| < d^*$ with $d^* > 0$, it is derived from (1.5) that

$$|e_y(t)| \le e^{-(a+k)t}|e_y(0)| + \frac{d^*}{a+k}\left[1 - e^{-(a+k)t}\right]. \tag{1.6}$$

Taking the limits of both sides of inequality, Equation (1.6) gives

$$|e_y(\infty)| \le \frac{d^*}{a+k}. \tag{1.7}$$

Particularly, suppose that the disturbance $d(t)$ is a constant one, that is, $d(t) = d_c$. It is derived from (1.5) that

$$e_y(t) = e^{-(a+k)t}e_y(0) - \frac{1}{a+k}\left[1 - e^{-(a+k)t}\right]d_c, \tag{1.8}$$

and consequently,

$$e_y(\infty) = -\frac{d_c}{a+k} \tag{1.9}$$

It is concluded from both (1.7) and (1.9) that proportional control can not completely remove the effects caused by disturbance (even a constant one) from the systems. In the presence of disturbance, to maintain a smaller tracking offset, a higher control gain k has to be designed to suppress the disturbance. This is the potential philosophy for high-gain control to reject the disturbances. Actually, this is the reason why many advanced feedback-control methods, such as H_∞ control and robust control, have been employed for disturbance rejection.

The performance of high-gain control to reject disturbance is evaluated by simulation studies of system (1.1) with parameter $a = 0.7$. The reference signal is taken as a step one $y_r(t) = 2$. A step disturbance $d(t) = 2$ is considered and imposed on the system at $t = 1$ sec. Three different control gains, $k = 6$, 20, and 200, respectively, are implemented in order to verify the disturbance rejection performance by high gain control, where the output and input responses are shown by Figure 1.2(a) and (b), respectively.

As shown by Figure 1.2(a), a higher control gain usually generates a faster tracking response and a smaller offset in the presence of disturbances. However, it can be

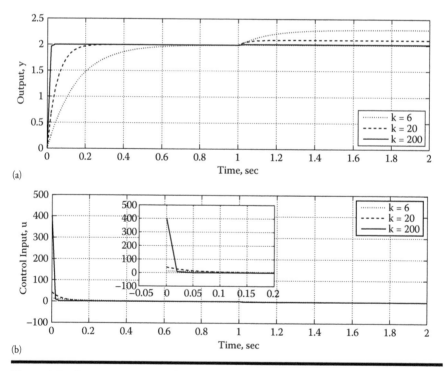

Figure 1.2 Response curves of system (1.1) in the presence of disturbance under the high gain control: (a) output; (b) control input.

observed from Figure 1.2(b) that a rather higher control energy is generally required so as to achieve a faster tracking and disturbance rejection performance. The results of this simulation scenario demonstrate that *there exists a contradiction between disturbance rejection and reasonable control energy for the high gain control method.*

1.2.2 Integral Control

In practical engineering systems, the integral control action is always employed to remove the offset in the presence of disturbances and uncertainties. The integral control law is usually designed as

$$u = k_1 e_y + k_0 \int_0^t e_y(\tau)d\tau + ay_r, \tag{1.10}$$

where k_1 and k_0 are the proportional and integral coefficients to be designed, respectively.

The block diagram of a closed-loop system (1.1) under proportional control (1.4) is shown by Figure 1.3. Substituting control law (1.4) into tracking error system (1.2), yields

$$\dot{e}_y = -(a + k_1)e_y - k_0 \int_0^t e_y(\tau)d\tau - d. \tag{1.11}$$

Taking derivatives of both sides of (1.11) gives

$$\ddot{e}_y + (a + k_1)\dot{e}_y + k_0 e_y + \dot{d} = 0. \tag{1.12}$$

Let $e = [e_y, \ \dot{e}_y]^T$, it follows from (1.12) that

$$e(t) = e^{At}e(0) + \int_0^t e^{A(t-\tau)}b\dot{d}(\tau)d\tau, \tag{1.13}$$

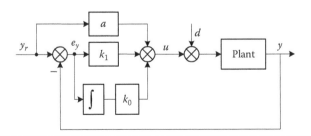

Figure 1.3   Block diagram of system (1.1) under the integral control.

where

$$A = \begin{bmatrix} 0 & 1 \\ -k_0 & -(a + k_1) \end{bmatrix}, \quad b = \begin{bmatrix} 0 \\ -1 \end{bmatrix}.$$

The control parameters should satisfy $k_0 > 0$ and $a + k_1 > 0$ (i.e., A is a Hurwitz matrix) so as to achieve asymptotical stability of the closed-loop system.

In the presence of constant disturbance, i.e., $\dot{d}(t) = 0$, since A is a Hurwitz matrix, it is derived from (1.13) that $e(\infty) = 0$ or equivalently $e_y(\infty) = \dot{e}_y(\infty) = 0$. This means that the integral control can finally remove the effects caused by constant disturbance from the system. However, the integral control can not remove the effects caused by nonconstant disturbances, such as harmonic ones. Suppose that the disturbance in (1.1) is nonconstant, but bounded and with a bounded derivative, i.e., $|\dot{d}(t)| < d^*$ with $d^* > 0$. It is derived from (1.13) that

$$\|e(t)\| \leq e^{At}\|e(0)\| + \left\| \int_0^t e^{A(t-\tau)} b\dot{d}(\tau) d\tau \right\|$$

$$\leq e^{At}\|e(0)\| + \left\| \int_0^t e^{A(t-\tau)} d\tau\, b \right\| \cdot d^* \qquad (1.14)$$

$$\leq e^{At}\|e(0)\| + \|A^{-1}\left(I - e^{At}\right) b\| \cdot d^*$$

Taking limits of both sides of inequality (1.6) yields

$$\|e(\infty)\| \leq \|A^{-1} b\| d^*. \qquad (1.15)$$

It is concluded from (1.15) that integral control can only remove the offset caused by constant disturbance. However, in the presence of nonconstant disturbances, the integral control may result in a steady-state tracking error. It is also noticed that the integral action in (1.4) always causes undesirable transient control performances, such as larger overshoot and longer settling time.

Simulation studies of system (1.1) under the integral control have been carried out to demonstrate its disturbance rejection performance. The parameter a of the system (1.1) is taken as $a = 0.7$. The reference signal is still taken as a step one $y_r(t) = 2$. The controller parameters are selected as $k_0 = 6$ and $k_1 = 15$. First, a step disturbance $d(t) = 2$ is considered and imposed on the system at $t = 10$ sec. And then, a harmonic disturbance $d(t) = 2 + 2\sin(t)$ is added to the system at $t = 20$ sec. The output and input response curves are shown by Figure 1.4(a) and (b), respectively.

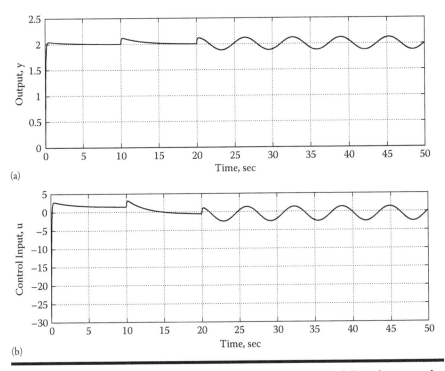

Figure 1.4 Response curves of system (1.1) in the presence of disturbance under the integral control: (a) output; (b) control input.

As shown by Figure 1.4(a), the integral control results in additional overshoot and settling time for the transient tracking process. In the presence of constant disturbance, the integral control can remove the offset of the tracking error. However, as shown by Figure 1.4(a), the integral control is unavailable to reject the harmonic disturbance. The results of this simulation scenario demonstrate that *there is a contradiction between disturbance rejection and tracking performance for the integral control method*. In addition, *the integral control could effectively reject the constant disturbance but achieves poor performance in the presence of time-varying disturbances, such as harmonic ones.*

1.2.3 Disturbance Observer-Based Control

As discussed earlier in the last section, the high gain control and integral control can be recognized as a PADC method. Different from the high gain control and integral control approaches, disturbance observer-based control (DOBC) provides an active and effective way to handle disturbances and improve robustness of the closed-loop

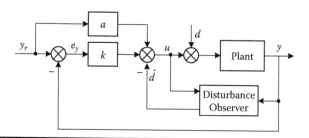

Figure 1.5 Block diagram of system (1.1) under the DOBC method.

system, which is always considered as an AADC method. The DOBC law is usually designed as

$$u = ke_y - \hat{d} + ay_r, \tag{1.16}$$

where k is the feedback control gain, and $\hat{d}$ is the disturbance estimation by a disturbance observer.

The block diagram of closed-loop system (1.1) under DOBC (1.6) is shown by Figure 1.5. Substituting control law (1.6) into tracking error system (1.2) yields

$$\dot{e}_y = -(a + k)e_y + e_d, \tag{1.17}$$

where $e_d = \hat{d} - d$ is the disturbance estimation error. By designing an appropriate disturbance observer, the disturbance estimation error e_d can be usually governed by

$$\dot{e}_d = f(e_d), \tag{1.18}$$

which is globally asymptotically stable.

The closed-loop systems consisting of tracking error dynamics (1.17) and observer error dynamics (1.18) can be considered as a cascaded system. It follows from the stability criterion that the cascaded system is asymptotically stable if the control gain is designed to satisfy $a + k > 0$. This implies that the disturbances can be finally suppressed asymptotically. Compared with the integral control method, the DOBC can compensate not only constant disturbances but also many other types of disturbances, such as harmonic ones, as long as they can be accurately estimated by a disturbance observer.

Simulation studies of system (1.1) under the DOBC method have been carried out to evaluate its effectiveness in disturbance rejection. The parameter a of system (1.1) is taken the same as the previous simulation scenarios, i.e., $a = 0.7$. The reference signal is still taken as a step one $y_r(t) = 2$. A step disturbance $d(t) = 2$ is imposed on the system at $t = 10$ sec, and a harmonic disturbance $d(t) = 2 + 2\sin(t)$ is imposed on the system at $t = 20$ sec. The feedback control parameter of DOBC is selected as $k = 6$. A disturbance observer

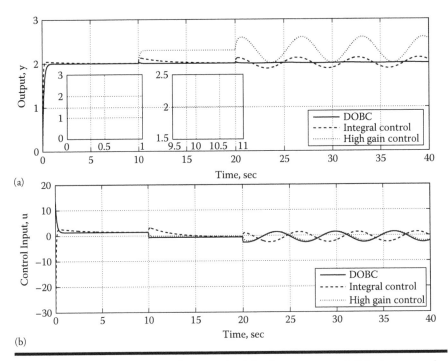

(a)

(b)

Figure 1.6   **Response curves of system (1.1) in the presence of disturbance under the DOBC method: (a) output; (b) control input.**

is employed to estimate such disturbances timely. To show the superiority of the DOBC method in disturbance rejection, both the proportional control and the integral control methods are implemented here for comparison studies. The control gain of the proportional control is chosen the same as the DOBC method. The control parameters of the integral control are selected as $k_0 = 6$ and $k_1 = 15$. The output and input response curves are shown by Figure 1.6(a) and (b), respectively.

As shown by Figure 1.6(a), the DOBC method could reject both constant and harmonic disturbances more promptly as compared with both the proportional control and the integral control methods, and no excessive control energy is required for disturbance rejection. In addition, it is observed from Figure 1.6 that the response curves under the DOBC method are overlapped with those of the baseline proportional control method, which is called nominal performance recovery. The reason for this lies in that the disturbance observer-based compensation serves like a patch to the baseline proportional control. In the absence of disturbances, only zero disturbance estimate is outputted by the disturbance observer, and the DOBC law reduces to the baseline proportional control law. In this case, no undesirable transient performance and excessive control energy are required in order to reject the disturbances. Generally speaking, *DOBC could achieve a good disturbance-rejection performance without scarifying the nominal performance.*

1.3 Basic Framework

A basic framework of disturbance observer based control method is shown in Figure 1.7. As shown in this figure, the composite controller consists of two parts: a feedback control part and a feedforward control part based on a disturbance observer. The feedback control is generally employed for tracking and stabilization of the nominal dynamics of the controlled plant. In this stage, the disturbances and uncertainties are not necessarily required to be considered. The disturbances and uncertainties on controlled plant are estimated by a disturbance observer and then compensated by a feedforward control. The major merit of such design lies in that the feedback control and feedforward designs satisfy the so-called separation principal, that is, the tracking control performance and the disturbance rejection performance can be achieved by adjusting the feedback and feedforward controllers, respectively. Such a promising feature has induced many superiorities as compared with the PADC methods, which are stated as follows:

- **Faster responses in handfing the disturbances**: As compared with the PADC method, DOBC always achieves a faster dynamic response in handling disturbances. The reason is that DOBC provides a feedforward compensation term to directly counteract the disturbances in the control systems, while the PADC only rejects the disturbances by passive feedback regulation.
- **"Patch" features**: The disturbance feedforward compensation term can be considered as a patch to the existing feedback control. The benefit of this is that there is no change to the baseline control, which may have been widely used and developed for many years, such as gain scheduling feedback control for classical flight control systems, PID feedback control for industrial systems, model predictive feedback control for process systems, etc. After the baseline feedback control is designed by using the conventional feedback control techniques, the DO-based compensation is added to improve the robustness and disturbance

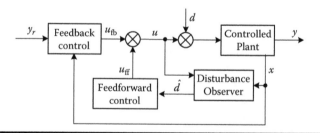

Figure 1.7   A basic framework of disturbance observer based control.

attenuation abilities. Instead of employing a completely new and different control strategy that demands a new verification and certification process, the verification of DOBC can be developed based on the existing verification process to ensure safety and reliability.

■ **Less conservativeness**: DOBC is not a worst-case-based design that can estimate and compensate disturbances online, thus has prominent adaptiveness and low conservativeness. Most of the existing robust control methods are worst-case-based design, where promising robustness is achieved with the price of degraded nominal performance, and thus have been criticized as being over-conservative. In DOBC approach, the nominal performance of the baseline controller is recovered in the absence of disturbances or uncertainties, thus a better nominal dynamic performance would be achieved.

1.3.1 Frequency Domain Formulation

For the sake of simplicity, let us consider a minimum-phase single-input, single-output linear system, depicted by the following frequency domain form

$$Y(s) = G_p(s)[U(s) + D(s)], \tag{1.19}$$

where $U(s)$ the control input, $Y(s)$ the controlled output, $D(s)$ the disturbance, and $G_p(s)$ the model of plant. Here, the disturbance is supposed to be a low-frequency one, which implies that $|d(j\omega)|$ is bounded for a low, frequency region $0 < \omega < \omega^*$, where ω^* is a small constant.

The block diagram of the DOBC for system (1.19) under the frequency domain description is given by Figure 1.8 [27], where $G_n(s)$ the nominal model of plant, $Q(s)$ the filter of disturbance observer, and $C(s)$ the feedback control part. The input–output representation of system (1.19) under the frequency domain DOBC is expressed as follows.

$$Y(s) = T_{yr}(s)Y_r(s) + T_{yd}(s)D(s), \tag{1.20}$$

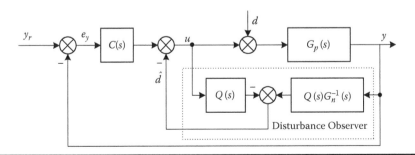

Figure 1.8 Block diagram of DOBC under the frequency domain formulation.

where

$$T_{ry}(s) = \frac{G_n(s)G_p(s)C(s)}{G_n(s)[1 + C(s)G_p(s)] + Q(s)[G_p(s) - G_n(s)]}, \tag{1.21}$$

$$T_{dy}(s) = \frac{G_n(s)G_p(s)[1 - Q(s)]}{G_n(s)[1 + C(s)G_p(s)] + Q(s)[G_p(s) - G_n(s)]}. \tag{1.22}$$

It follows from (1.20)–(1.22) that the disturbance rejection property mainly depends on the design of filter $Q(s)$ in the frequency domain-disturbance observer. Actually, if $Q(s)$ is chosen as a low-pass filter with a steady-state gain of 1, i.e., $\lim_{\omega \to 0} Q(j\omega) = 1$, it can be obtained from (1.22) that

$$\lim_{\omega \to 0} T_{ry}(j\omega) = \lim_{\omega \to 0} \frac{G_n(j\omega)C(j\omega)}{1 + G_n(j\omega)C(j\omega)}, \tag{1.23}$$

and

$$\lim_{\omega \to 0} T_{dy}(j\omega) = 0. \tag{1.24}$$

Supposing that the transfer functions in (1.21) and (1.22) are stable, then Equation (1.23) implies that the real uncertain closed-loop system under the frequency domain DOBC behaves as if it is the nominal closed-loop system in the absence of disturbance, which is also referred to as the nominal performance recovery. It follows from (1.24) that the low-frequency disturbances can be attenuated asymptotically under the frequency domain DOBC.

1.3.2 Time Domain Formulation

Consider a multi-input multi-output linear system with disturbances, depicted by

$$\begin{cases} \dot{x} = Ax + B(u + d), \\ y = Cx, \end{cases} \tag{1.25}$$

where $x \in R^n$, $u \in R^m$, $d \in R^m$, and $y \in R^l$ represent state, control input, disturbance, and output vectors, A, B, and C represent system matrices with dimensions of $n \times n$, $n \times m$, and $l \times n$, respectively.

For the sake of simplicity, suppose that the disturbances and their derivatives are bounded and tend to some constants as time goes to infinity, that is, $\lim_{t \to \infty} \dot{d}(t) = 0$ or $\lim_{t \to \infty} d(t) = d_s$, where d_s is a constant vector. The following time domain disturbance observer can be employed to estimate the disturbances in system (1.25), given by [2]

$$\begin{cases} \dot{z} = -LB(z + Lx) - L(Ax + Bu), \\ \hat{d} = z + Lx, \end{cases} \tag{1.26}$$

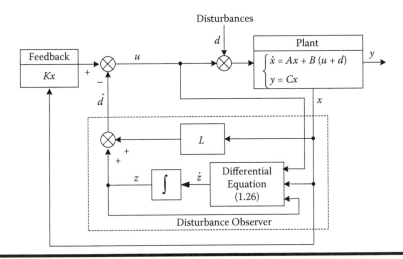

Figure 1.9   Block diagram of DOBC under the time domain formulation.

where $\hat{d}$ the disturbance estimation vector, z the internal variable vector of the observer, and L the observe gain matrix to be designed.

Under the time domain framework of DOBC, the control law for system (1.25) is generally designed as

$$u = Kx - \hat{d}, \tag{1.27}$$

where K is the feedback control law to be designed. The block diagram of the DOBC for system (1.25) under the time domain description is given by Figure 1.9.

The disturbance estimation error is defined as

$$e_d = \hat{d} - d. \tag{1.28}$$

Combining Equations (1.25)–(1.28), the closed-loop system is governed by

$$\begin{cases} \dot{x} = (A + BK)x - Be_d, \\ \dot{e}_d = -LBe_d - \dot{d}. \end{cases} \tag{1.29}$$

It can be shown that closed-loop system (1.29) is bounded-input, bounded-output (BIBO) stable if the feedback control gain is selected such that $A + BK$ is Hurwitz, and the observer gain matrix is selected such that $-LB$ is Hurwitz. In addition, if the disturbances tend to constants, it can be shown that closed-loop system (1.29) is asymptotically stable with appropriately chosen parameters K and L, such that $A + BK$ and $-LB$ are Hurwitz.

1.4 Early History

1.4.1 An Overview on Disturbance Estimation Approaches

Since 1970s, many effective disturbance estimation techniques have been developed, such as Unknown Input Observer (UIO) [69], Perturbation Observer (PO) [70, 71], Equivalent Input Disturbance (EID)-based estimator [72, 73], Extended State Observer (ESO) [74, 75, 4, 76], and Disturbance Observer (DO) [2, 27, 26, 28]. Among those disturbance estimation approaches, the theory and applications of DO and ESO are most extensively investigated. ESO was firstly proposed by Prof. J. Han in the 1990s [9]. ESO is generally regarded as a fundamental part of the so-called active disturbance rejection control (ADRC), which is often employed to estimate the lumped disturbances consisting of unknown uncertainties and external disturbances. DO was initiatively put forward by Prof. K. Ohishi and his colleagues in the late 1980s [27]. The analysis and design method for DO has achieved significant progress during the past three decades. On the basis of conventional frequency-domain linear DO, the developments of nonlinear DO and intelligent DO techniques by using nonlinear system theory and intelligent system theory has become very active. The research status of DO may be illustrated as follows.

1.4.1.1 Linear Disturbance Observer

Due to the perfection and maturity of linear system theory, linear disturbance observer technique has obtained great progress during the past three decades, and the stability or robust stability analysis of LDO is always one of the hottest topics in control society. In [28], a LDO was employed to estimate the torque disturbance of a DC servo speed control system, where the robustness of system is analyzed by using frequency-domain theory. In [77], a robust LDO was presented for a class of minimum-phase systems, where the LDO filter selection criterion for the second-order minimum phase systems was provided. An almost necessary and sufficient condition for robust stability of the closed-loop systems with LDO was presented for minimum-phase systems in [78], which enlarges the region of parameter selection of the LDO filter.

Note that the formation and the parameter selection of LDO filter has a significant impact on control system performance, which has been widely investigated in literature. Ohishi et al. have selected the LDO filter with a first-order low-pass formation [27], which has a straightforward system structure and easy tuning parameters but is only available for the first-order system. [26] has suggested to fix the relative order of the filter in LDO according to the practical plant, while the orders of numerator and denominator can be increased relatively. Such an approach enriches the frequency features of a system and would achieve a better disturbance estimation performance. However, there are too many parameters that are not easy to tune. In [79], a high-order filter in LDO was proposed to attenuate fast time-varying disturbances and has achieved promising performances. However, the existence of a large phase lag in the filter deteriorated its damping features. As analyzed above, the filter selection of LDO should make a compromise among a variety of factors, such as stability, disturbance rejection performance, and difficulties of parameter tuning. To this end, it is especially important to select particular structure and parameters of the filter for a concrete plant to achieve a satisfactory performance.

In addition, the LDO is usually employed for nonlinear systems to estimate the lumped disturbances that consist of external disturbances and also nonlinear terms that are difficult to be handled by feedback control design. In [80], for a class of nonlinear systems, a LDO was designed to estimate the nonlinear terms that can be divided into linear time invariant parts and bounded nonlinear parts.

In conclusion, the LDO has a relatively mature theoretic foundation and intuitive and simple structure. It is available for not only linear systems but also some nonlinear systems with particular structures. The LDO has gained wide and successful applications in various industrial systems, such as manipulators [23, 24, 25], hard disk drives [38, 39], servo motors [27, 28, 29, 30, 31], and general motion control systems [81, 82, 83].

1.4.1.2 Nonlinear Disturbance Observer

Although it has been reported by many researchers that the nonlinear terms in weak nonlinear systems can be regarded as disturbances and LDOs can be used to estimate those nonlinearities [84, 12], the LDO generally requires these nonlinear systems to satisfy certain particular structure, and it is unavailable for more general nonlinear systems. To this end, the investigation of nonlinear disturbance observer (NDO) for essential nonlinear systems has become one of the hottest topics during the recent decade.

In [2], a NDO was proposed to estimate disturbance torques in nonlinear robotic manipulators, where it was shown by Lyapunov stability theory that the disturbance estimation error exponentially converges to zero. However, only constant disturbances could be handled in [2]. For a class of affine nonlinear systems subject to disturbances governed by exogenous system, a NDO was put forward to estimate the disturbances with an exponential convergence rate by constructing a nonlinear observer gain function in [5]. A NDO with sliding mode structure was proposed for minimum-phase (with respect to the relationship between disturbances and outputs) dynamical systems with arbitrary relative degrees in [85], where the control performance depends on the estimation of the upper and lower bounds of disturbances. A NDO was proposed in [6] to estimate harmonic disturbances for a class of single-input single-output nonlinear systems. In [86], a NDO was proposed for a class of multiinput multioutput nonlinear systems subject to disturbances governed by exogenous linear systems. Recently, a generalized NDO was presented in [11] to estimate high-order disturbances. In addition, by integrating the ideas of disturbance estimation and the intelligent control, intelligent disturbance observers are also widely investigated, such as fuzzy disturbance observer [87, 88] and neural network disturbance observer [89].

It can be concluded from the above analysis that the development of NDO is of great importance for disturbance estimation of nonlinear systems. However, the current research results are constrained to several particular forms of nonlinear systems with limited types of disturbances, such as constant and harmonic ones. It is imperative to develop new NDOs to estimate the disturbances with more general formulations for more general nonlinear systems.

1.4.2 An Overview of Disturbance Estimation-Based Control Approaches

Besides the disturbance estimation techniques, the design and analysis of the composite control approach would be another challenging but important topic. In this part, the major three topics regarding the composite control design and analysis are reviewed as follows.

1.4.2.1 Robustness Performance and Stability

One of the major reasons for the successful applications of DOBC in practical engineering lies in that it could effectively estimate and then compensate the lumped disturbances including model uncertainties, parameter perturbations, and unmodeled dynamics, as well as external disturbances, and thus achieves strong robustness. However, there still exist some open problems, for example, the analysis of robust stability and robustness performance for uncertain systems under DOBC.

First, in most existing results, it has been reported via plenty of experimental or application results that the DOBC approach could achieve very excellent dynamic and static performances in the presence of model uncertainties. However, it should be pointed out that DOBC is still a model-based control approach, and for a designed DOBC, it is hard to achieve fine control performance in the presence of severe model uncertainties. Actually, for a tuned DOBC, it can be verified by simulation studies that the control performance even the stability of closed-loop system is hard to be guaranteed if there exist enough large model uncertainties. To this end, the research problem that how large model uncertainties the DOBC can handle becomes a major concern.

Recently, several scholars have carried out some researches on this topic. In [90], the singular perturbation theory was employed for robust stability analysis for linear uncertain systems under DOBC. It has been pointed out there that the stability of closed-loop system was guaranteed for arbitrary model uncertainties as long as the bandwidth of filter in DO was designed widely enough. Similarly, it was derived in [78] by using frequency-domain internal stability that the minimum-phase linear systems under DOBC is stable as long as the time constant of filter was set to be smaller enough. These kinds of results are very important for the development of DOBC as they provided theory guidances for practical applications. However, the current results are limited to linear systems. The robust stability analysis on general nonlinear systems under DOBC would be more crucial but challenging.

Secondly, as discussed previously, one of the major merits of DOBC as compared with other robust control methods lies in its nominal performance recovery. However, in the presence of severe model uncertainties, the transient performance can not be guaranteed for the existing DOBC methods. By using the singular perturbation theory, a DOBC method with transient performance recovery was proposed for a class of single-input, single-output uncertain nonlinear systems in [91] and then extended to multi-input multi-output nonlinear systems in [92].

Generally speaking, the results on robust stability and robustness performance analysis of closed-loop system under DOBC are not enough presently, and further investigations are required to enrich the current theoretic results.

1.4.2.2 Composite Hierarchical Anti-Disturbance Control

Due to the limitations of pure feedforward compensation in disturbance rejection, such as the robust stability issue and the limitation of applicability, the introduction of advanced feedback control to the DOBC becomes another hot topic, which formulates the so-called composite hierarchical anti-disturbance control (CHADC) [86, 93, 94, 95, 96, 97, 98, 99]. The basic idea of CHADC can be explained as follows: the disturbances and uncertainties in control system is firstly stratified into two parts, where one part (usually includes disturbances) is feedforward compensated by DOBC while another part (usually includes plant uncertainties) is attenuated by advanced feedback control. The advantages lie in that the stability of closed-loop system is guaranteed with the composite hierarchical anti-disturbance control. For example, the DOBC is combined with H-infinity control in [93, 95], and with variable structure control in [94, 96, 99].

Although the CHADC enriches the theory and application research of the field of DOBC in different aspects, it still exhibits some application limitations. First, the types of exogenous disturbances are quite limited, such as unknown constant ones or harmonic ones with known frequency. Second, the uncertainties are generally required to satisfy some special assumptions such as with a bounded H_2 norm due to robust stability conditions of the H-infinity control and variable structure control.

1.4.2.3 Compensation of Mismatched Disturbances

Most of the existing DOBC approaches are generally confined to uncertain systems which satisfy the so-called matching conditions. Here, matching refers to the disturbances acting on system via the same channels as control inputs, or the disturbances can be transferred to the same channels as the control inputs by coordinate transformations [100]. However, the mismatched disturbances (i.e., the disturbances that enter the system through different channels from those of the control inputs) are more general cases (as compared with matched disturbances and uncertainties) and widely exist in practical applications. Taking an aircraft as an example, the lumped disturbance torques caused by unmodeled dynamics, external winds, and parameter perturbations may affect aircraft dynamics in a rather complicated way, which do not necessarily satisfy the so-called matching condition [40, 13]. The problem also appears in a permanent magnet synchronous motor (PMSM) system, where the uncertainties consisting of parameter variations and load torque enter the system via different channels from those of the voltage inputs [101, 102]. Another example is the MAGnetic LEViation (MAGLEV) suspension system, where the track input disturbances act on different channels from the control input [12, 103].

When confronted with mismatched disturbances and uncertainties, the most widely used practical method is to add an integral action in a feedback control law to remove offset of the closed-loop system. However, as analyzed previously, the integral control always brings many undesired control performances, such as longer settling time and larger overshoot. In addition, the robust control approaches are usually employed to handle the mismatched uncertainties. However, the robustness is obtained at the price of sacrificing the nominal control performance.

It is also worth noting that the problem of rejecting mismatched disturbances has been studied in [93, 97] where the matched disturbances are compensated by DOBC while the mismatched disturbances are attenuated by some traditional feedback controllers, such as variable structure control [97] and H-infinity control [93]. It is also reported that certain constraints in the mismatched disturbances (such as bounded H_2 norm) are required in [93, 97].

1.5 What Is in This Book

This book aims to provide an intuitive way for research scholars and also industrial practitioners to understand the main concepts and design philosophies of disturbance observer-based control method.

The contents of this book has been divided into six major parts. In the first part, an overview on the DOBC methods and applications has been given. Several practical and popular disturbance estimation techniques have been presented in Part 2. In Part 3, the design and analysis of DOBC methods have been addressed. Applications of the DOBC methods to process control systems, mechatronic systems, and flight control systems have been presented in Parts 4–6, respectively.

This chapter has been considered as the first part of the book, which has given an overview of DOBC methods and applications. The major motivation of the development of DOBC technique has been described via a numerical example. The basic frameworks and also the early history of DOBC methods are briefly reviewed in this chapter.

The disturbance estimation techniques are introduced in Part 2, which consists of three chapters. Chapter 2 has introduced the major linear disturbance estimation techniques, including frequency domain disturbance observer, time domain disturbance observer, and extended state observer. In Chapter 3, some traditional nonlinear disturbance observers for estimating constant and harmonic disturbances have been presented. Finally, three advanced nonlinear disturbance estimation techniques including high-order disturbance observer, extended high-gain state observer and finite-time disturbance observer have been given in Chapter 4 for further deep performance improvement in disturbance estimation.

Part 3 consists of five chapters, which mainly consider the disturbance compensation problem in the field of DOBC. In Chapter 5, a generalized framework for DOBC design is presented. In Chapter 6, a generalized extended state observer-based control method for compensating the mismatched disturbances in linear systems has been discussed by appropriately designing a linear disturbance compensation gain matrix. In Chapter 7, the results have been extended to nonlinear systems under mismatched disturbances, where a corresponding nonlinear disturbance compensation gain has been designed. With the help of the definition of differential geometry theory, the disturbance compensation problem for nonlinear systems with arbitrary relative degrees has been discussed in Chapter 8. Finally, in Chapter 9, the mismatched disturbance compensation problem in sliding mode control field has been solved by designing a dynamic sliding surface based on the estimation of disturbances by a disturbance observer.

In Part 4, the application design of DOBC for typical process control problems has been presented. Firstly, the application of DOBC method to a typical level tank control system has been provided with both simulation and experimental verifications in Chapter 10. In Chapter 11, the disturbance rejection problem of ball mill grinding circuits via DOBC has been given with detailed control design and simulation studies.

In Part 5, we are concerned with the application of DOBC for disturbance rejection in mechatronic systems fields. In Chapter 12, an adaptive control design for speed control of a permanent magnet synchronous motor system with disturbance rejection property has been considered. In Chapter 13, the application of DOBC to air gap control of a MAGLEV suspension system has been considered.

In the final part, the disturbance rejection for flight control systems has been discussed. In Chapter 14, the position control problem of miniature helicopters under disturbances has been considered by using a nonlinear disturbance observer, where the results have shown that the DOBC method significantly improves the disturbance-rejection performance. In Chapter 15, the attitude control of bank-to-turn missiles with severe disturbances and aerodynamic parameter perturbations has been solved by a DOBC method. Finally, in Chapter 16, the disturbance rejection of hypersonic vehicles under mismatched disturbances has been presented by a nonlinear DOBC method.

DISTURBANCE ESTIMATION DESIGN

Chapter 2

Linear Disturbance Estimator

2.1 Introduction

In this chapter, several commonly used linear disturbance estimation techniques are introduced. The frequency domain disturbance observer, which was proposed by [27] in the industrial application society in late 1980s, is presented first for both minimum phase and nonminimum phase cases. The time domain formulation of the disturbance observer is provided with a detailed analysis later. Finally, the extended state observer technique, which can simultaneously estimate states and disturbances, is presented.

2.2 Frequency Domain Disturbance Observer

2.2.1 Minimum-Phase Case

Consider a single-input single-output linear minimum phase system, depicted by the following frequency domain form

$$Y(s) = G_p(s)[U(s) + D(s)], \qquad (2.1)$$

where $U(s)$ the control input, $Y(s)$ the controlled output, $D(s)$ the disturbance, and $G_p(s)$ the model of plant.

The block diagram of the frequency domain disturbance observer for system (2.1) is given by Figure 2.1(a) [27], where $G_n(s)$ the nominal model of plant and $Q(s)$, the filter of disturbance observer, respectively. Note that the disturbance observer

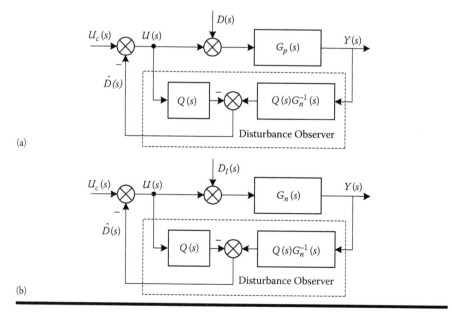

(a)

(b)

Figure 2.1 **Block diagram of the frequency-domain disturbance observer for minimum-phase linear system: (a) an original form; (b) an equivalent form.**

can estimate not only the external disturbances but also the internal disturbances caused by model uncertainties. To show how the disturbance observer estimates the lumped disturbances consisting of both the external and internal ones, an equivalent block diagram of the original block diagram in Figure 2.1(a) is given in Figure 2.1(b), where the lumped disturbances are denoted by

$$D_l(s) = G_n^{-1}(s) G_p(s) D(s) + [G_n^{-1}(s) G_p(s) - 1] U(s). \qquad (2.2)$$

It follows from Figure 2.1(b) that

$$\begin{aligned}
\hat{D}(s) &= Q(s) G_n^{-1}(s) Y(s) - Q(s) U(s) \\
&= Q(s) G_n^{-1}(s) G_n(s)(U(s) + D_l(s)) - Q(s) U(s) \qquad (2.3) \\
&= Q(s) D_l(s).
\end{aligned}$$

It is derived from (2.3) that

$$E_d(s) = \hat{D}(s) - D_l(s) = [Q(s) - 1] D_l(s). \qquad (2.4)$$

The lumped disturbance estimation error $E_d(s)$ will tend to zero as time goes to infinity if the filter $Q(s)$ is selected as a low-pass form, that is, $\lim_{s \to 0} Q(s) = 1$.

It is also derived from Figure 2.1(a) that the output can be represented as

$$Y(s) = G_{uy}(s)U_c(s) + G_{dy}(s)D(s). \tag{2.5}$$

where

$$G_{uy}(s) = \frac{G_p(s)G_n(s)}{G_n(s) + Q(s)[G_p(s) - G_n(s)]}, \tag{2.6}$$

$$G_{dy}(s) = \frac{G_p(s)G_n(s)[1 - Q(s)]}{G_n(s) + Q(s)[G_p(s) - G_n(s)]}. \tag{2.7}$$

Clearly, if the filter $Q(s)$ is selected as a low-pass form, i.e., $\lim_{\omega \to 0} Q(j\omega) = 1$, it follows from (2.6) and (2.7) that

$$\lim_{\omega \to 0} G_{uy}(j\omega) = G_n(j\omega), \tag{2.8}$$

and

$$\lim_{\omega \to 0} G_{dy}(j\omega) = 0. \tag{2.9}$$

Equation (2.8) implies that the system with the frequency-domain disturbance observer behaves as if it were the nominal plant in the low-frequency domain. It can be concluded from (2.9) that the low-frequency-domain disturbances have been eliminated from the system by feedforward compensation.

It follows from (2.4) and (2.20) that the performance of disturbance estimation is mainly determined by the design of low-pass filter $Q(s)$. Actually, $Q(s)$ has to be designed such that:

■ The relative degree of $Q(s)$, that is, the order difference between the denominator and the numerator, should be no less than that of the nominal model $G_n(s)$. This design principal is to make sure that the control structure is realizable, i.e., $Q(s)G_n^{-1}(s)$ should be proper;
■ In the domain of low-frequency, $Q(s)$ approaches to 1, guaranteeing that the estimate of lumped disturbance approximately equals to the lumped disturbance. This means that the effects of the disturbance can be attenuated by a feedforward compensation design based on DO.

The design and tuning principal of filter $Q(s)$ will be illustrated by the following numerical example. Consider a SISO minimum-phase system as stated by (2.1), where its nominal and plant models are given by

$$G_n(s) = \frac{s + 1}{(s + 0.5)(s + 2)}, \quad G_p(s) = \frac{s + 3}{(s + 1)(s + 4)}. \tag{2.10}$$

According to the above design guidelines, the filter $Q(s)$ can be selected as a first-order low-pass form with a steady-state gain of 1, i.e.,

$$Q(s) = \frac{1}{\lambda s + 1}. \tag{2.11}$$

In this case, the disturbance estimation accuracy depends on the selection of the filter parameter λ in $Q(s)$. Actually, it follows from (2.4) and (2.20) that the property of disturbance estimation is determined by the frequency characteristics of transfer function $1 - Q(s)$. The bode diagram of transfer function $1 - Q(s)$ under different filter parameters is shown by Figure 2.2.

As shown by Figure 2.2, the smaller the filter parameter λ is, the smaller the magnitude of transfer function $1 - Q(s)$ is. This implies that the estimation error of disturbance converges in a much smaller area by choosing a relatively smaller parameter λ in filter $Q(s)$.

In this simulation scenario, the external disturbance is taken as follows

$$d(t) = \begin{cases} \sin t, & 0 \le t \le 1, \\ 1 + \sin t, & t > 1. \end{cases}$$

The lumped disturbances consisting of external disturbances and internal disturbances caused by model uncertainties have been represented by (2.2). The response curves of lumped disturbance estimation for system (2.10) under different filter parameters are shown in Figure 2.3.

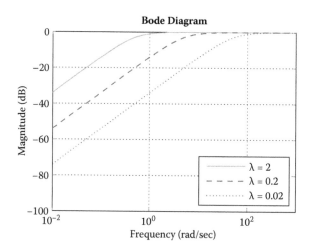

Figure 2.2 Bode diagram of transfer function $1 - Q(s)$ under different filter parameters.

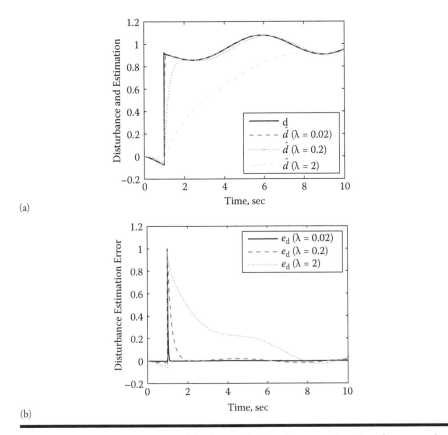

(a)

(b)

Figure 2.3 **Response curves of the lumped disturbance estimation for numerical example (2.10) under different filter parameters: (a) lumped disturbances and estimations; (b) estimation errors.**

It can be observed from Figure 2.3 that a smaller filter parameter λ has brought a better transient dynamics in estimation and a smaller static disturbance estimation error, while a lager filter parameter has resulted in a larger disturbance estimation error. Such simulation results precisely match the results of the above frequency domain analysis.

2.2.2 Nonminimum Phase Case

Consider the following transfer function

$$G_n(s) = \frac{k_p(-\beta s + 1)}{(\tau_{p1}s + 1)(\tau_{p2}s + 1)}e^{-\tau s}, \tag{2.12}$$

where τ_{p1}, τ_{p2}, and β are positive real numbers, indicating a right-half-plane (RHP) zero in the transfer function. Suppose that the filter in the disturbance observer is chosen as a first-order low-pass form

$$Q(s) = \frac{1}{\lambda s + 1}. \qquad (2.13)$$

If we use the previous method for the minimum phase case to construct the disturbance observer for system (2.12), it yields

$$G_n^{-1}(s)\, Q(s) = \frac{(\tau_{p1} s + 1)(\tau_{p2} s + 1)}{k_p(-\beta s + 1)(\lambda s + 1)} e^{\tau s}. \qquad (2.14)$$

Note that the zeros of the process model become the poles of disturbance observer. This results in an unstable observer and thus the possibility of unbounded disturbance estimation. To this end, if a process has RHP zeros, the process has to be factored out before using the model inverse for the observer design. There are a number of approaches to factor the same transfer function [104]. The most widely used method would be all-pass factorization, which places the RHP zero in the non-invertible part of the process model, and it also places a pole at the reflection of the RHP zero. Taking system (2.12) as an example, it is factored as follows:

$$G_n(s) = G_{n-}(s)\, G_{n+}(s), \qquad (2.15)$$

where

$$G_{n-}(s) = \frac{k_p(\beta s + 1)}{(\tau_{p1} s + 1)(\tau_{p2} s + 1)}, \qquad G_{n+}(s) = \frac{-\beta s + 1}{\beta s + 1} e^{-\tau s}.$$

The steady state gain of $G_{n+}(s)$ is always one for such factorization method. The block diagram of the frequency-domain disturbance observer for nonminimum phase linear system is shown in Figure 2.4 [105].

It is obtained from Figure 2.4 that the output is represented as

$$Y(s) = G_{uy}(s)\, U_c(s) + G_{dy}(s)\, D(s). \qquad (2.16)$$

where

$$G_{uy}(s) = \frac{G_{n-}(s)\, G_p(s)}{G_{n-}(s) + Q(s)[G_p(s) - G_n(s)]}, \qquad (2.17)$$

$$G_{dy}(s) = \frac{G_{n-}(s)[1 - G_{n+}(s)\, Q(s)]\, G_p(s)}{G_{n-}(s) + Q(s)[G_p(s) - G_n(s)]}. \qquad (2.18)$$

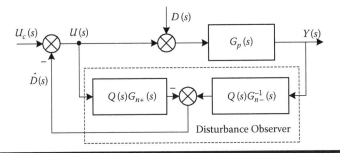

Figure 2.4   Block diagram of the frequency-domain disturbance observer for non-minimum-phase linear system.

Note that the factorization ensures that $\lim_{\omega \to 0} G_{n+}(j\omega) = 1$ and also the filter $Q(s)$ is selected as a low-pass form, i.e., $\lim_{\omega \to 0} Q(j\omega) = 1$, it follows from (2.17) and (2.18) that

$$\lim_{\omega \to 0} G_{uy}(j\omega) = G_{n-}(j\omega), \tag{2.19}$$

and

$$\lim_{\omega \to 0} G_{dy}(j\omega) = 0. \tag{2.20}$$

Similarly, Equation (2.19) implies that the system with frequency domain disturbance observer behaves as if it were the nominal plant $G_{n-}(s)$ in the low frequency domain. It can be concluded from (2.20) that the low-frequency-domain disturbances for such nonminimum phase system have been eliminated from the system by feedforward compensation.

Suppose the parameters in system (2.12) are taken as

$$k_p = 0.8, \quad \beta = 0.1, \quad \tau_{p1} = 3, \quad \tau_{p2} = 1.5, \quad \tau = 0.5.$$

For the sake of simplicity, the disturbance is taken as a step one: $d(t) = 3$ for $t \geq 2$ sec. The response curves of the disturbance estimations under different filter parameters are shown in Figure 2.5.

As shown in Figure 2.5, the disturbances can be estimated without offset under three different filter parameters. It is observed from the figure that the smaller the filter parameter is, the faster the convergence rate of estimation error is. However, it is also noticed that a smaller filter parameter will result in a larger nonminimum phase effects of error dynamics of disturbance observer. In conclusion, there exists a contradiction between faster disturbance-estimation dynamics and smaller nonminimum phase effects. It is the users' right to choose an appropriate filter parameter

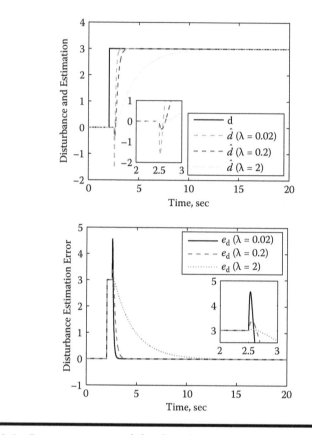

(a)

(b)

Figure 2.5 **Response curves of the disturbance estimation for numerical example (2.12) under different filter parameters: (a) disturbances and estimations; (b) estimation errors.**

to make a balance between the observer dynamics and the nonminimum phase effects.

2.3 Time Domain Disturbance Observer

Consider a multi-input multi-output linear system with disturbances, depicted by

$$\begin{cases} \dot{x} = Ax + B_u u + B_d d, \\ y = Cx, \end{cases} \tag{2.21}$$

where $x \in R^n$, $u \in R^m$, $d \in R^r$, and $y \in R^l$ represent state, control input, disturbance, and output vectors. A, B_u, B_d, and C represent system matrices with dimensions of $n \times n$, $n \times m$, $n \times r$, and $l \times n$, respectively.

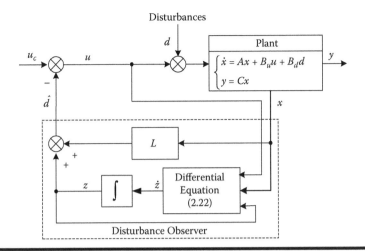

Figure 2.6 **Block diagram of the time-domain disturbance observer for linear system.**

The following time-domain disturbance observer can be employed to estimate the disturbances in system (2.21), given by [13]

$$\begin{cases} \dot{z} = -L B_d(z + Lx) - L(Ax + B_u u), \\ \hat{d} = z + Lx, \end{cases} \tag{2.22}$$

where $\hat{d}$ the disturbance estimation vector, z the internal variable vector of the observer, and L the observer gain matrix to be designed.

The block diagram of the time-domain disturbance observer for system (2.21) is given by Figure 2.6.

The disturbance estimation error is defined as

$$e_d = \hat{d} - d. \tag{2.23}$$

Taking the derivative of the disturbance estimation error in (2.23) along with system dynamics (2.21) and observer dynamics (2.22) gives

$$\dot{e}_d = -L B_d e_d - \dot{d}. \tag{2.24}$$

It is shown that the disturbance estimation error system (2.24) is BIBO stable if the observer gain matrix is chosen such that $-LB_d$ is Hurwitz. In addition, if the disturbances tend to constants, it is also shown that estimation error system (2.24) is asymptotically stable with appropriately chosen parameter L such that $-LB_d$ is Hurwitz.

Consider linear system (2.21) with matrices taken as

$$A = \begin{bmatrix} -0.8 & 1 & 1.6 \\ 0 & -3 & 2 \\ 0 & 0 & -6 \end{bmatrix}, \quad B_u = \begin{bmatrix} 0 & 0 \\ 1 & 0 \\ 0 & 1 \end{bmatrix}, \quad B_d = \begin{bmatrix} 0.8 & 0 \\ 0 & -1 \\ -0.4 & 1.2 \end{bmatrix}.$$

The gain matrix of disturbance observer (2.22) is designed as

$$L = \begin{bmatrix} 40 & 0 & -20 \\ 0 & -20 & 60 \end{bmatrix}.$$

It can be calculated from the above setting that

$$-LB_d = \begin{bmatrix} -40 & 24 \\ 24 & -92 \end{bmatrix}.$$

The eigenvalues of matrix $-LB_d$ are calculated as $\lambda_1 = -101.3836$ and $\lambda_2 = -30.6164$, which implies that matrix $-LB_d$ is Hurwitz and satisfies the design guidelines stated above. The response curves of disturbance estimations under different filter parameters are shown in Figure 2.7.

It is observed from Figure 2.7 that the time-domain disturbance observer (2.22) with the above designed parameters has achieved quite fine disturbance estimation performances. Note that the time-domain DO (2.22) here can be used for both minimum and nonminimum phase MIMO linear systems. However, it requires all

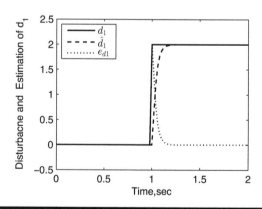

(a)

Figure 2.7 Response curves of the disturbance estimation for numerical example (2.21): (a) first disturbance d_1; (b) second disturbance d_2. *(Continued)*

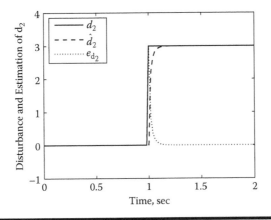

(b)

Figure 2.7 *(Continued)* **Response curves of the disturbance estimation for numerical example (2.21): (a) first disturbance d_1; (b) second disturbance d_2.**

the state information for observer design, while the frequency-domain DO only uses the output and input information.

2.4 Extended State Observer

Consider a class of SISO uncertain systems with order of n, described by the following differential equation

$$y^{(n)}(t) = f(y(t), \dot{y}(t), \ldots, y^{(n-1)}(t), d(t), t) + bu(t), \qquad (2.25)$$

where $d(t)$ the external disturbance, $u(t)$ the control input, $y(t)$ the controlled output, b a system parameter, and $f(y(t), \dot{y}(t), \ldots, y^{(n-1)}(t), d(t), t)$ the lumped disturbances consisting of the external one $d(t)$ and internal ones caused by model uncertainties.

Let $x_1 = y$, $x_2 = \dot{y}$, $x_3 = \ddot{y}$, ..., $x_n = y^{(n-1)}$, uncertain system (2.25) can be equivalently represented as

$$\begin{cases} \dot{x}_1 = x_2, \\ \dot{x}_2 = x_3, \\ \vdots \\ \dot{x}_{n-1} = x_n, \\ \dot{x}_n = f(x_1, x_2, \ldots, x_n, d(t), t) + bu, \\ y = x_1. \end{cases} \qquad (2.26)$$

The following augmented variable is usually introduced in the framework of an extended-state observer (ESO)

$$x_{n+1} = f(x_1, x_2, \ldots, x_n, d(t), t). \tag{2.27}$$

to linearize system (2.26). Combining (2.26) with (2.27), the extended-state equation is given by

$$
\begin{cases}
\dot{x}_1 = x_2, \\
\dot{x}_2 = x_3, \\
\quad \vdots \\
\dot{x}_{n-1} = x_n, \\
\dot{x}_n = x_{n+1} + bu, \\
\dot{x}_{n+1} = h(t), \\
y = x_1.
\end{cases}
\tag{2.28}
$$

where $h(t) = \dot{f}(x_1, x_2, \ldots, x_n, d(t), t)$.

In order to estimate the extended states, a linear ESO is generally designed as [9]

$$
\begin{cases}
\dot{z}_1 = z_2 - \beta_1(z_1 - y), \\
\dot{z}_2 = z_3 - \beta_2(z_1 - y), \\
\quad \vdots \\
\dot{z}_n = z_{n+1} - \beta_n(z_1 - y) + bu, \\
\dot{z}_{n+1} = -\beta_{n+1}(z_1 - y),
\end{cases}
\tag{2.29}
$$

where $z_1, z_2, \ldots, z_n$, and z_{n+1} are estimates of states $x_1, x_2, \ldots, x_n$, and x_{n+1}, respectively, and $\beta_1, \beta_2, \ldots, \beta_{n+1}$ are the observer gains to be designed. The block diagram of the ESO for system (2.26) is shown in Figure 2.8.

Subtracting (2.28) from (2.29), the observer error system is governed by

$$
\begin{cases}
\dot{e}_1 = e_2 - \beta_1 e_1, \\
\dot{e}_2 = e_3 - \beta_2 e_1, \\
\quad \vdots \\
\dot{e}_n = e_{n+1} - \beta_n e_1, \\
\dot{e}_{n+1} = -\beta_{n+1} e_1 - h(t),
\end{cases}
\tag{2.30}
$$

where $e_i = z_i - x_i$ $(i = 1, 2, \ldots, n+1)$ represents the state estimation error. It has been shown that the BIBO stability of error system (2.30) is guaranteed under the assumption that $h(t)$ is bounded.

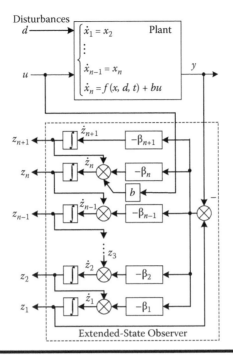

Figure 2.8 **Block diagram of the ESO for the linear system.**

Consider the following second-order system

$$\begin{cases} \dot{x}_1 = x_2, \\ \dot{x}_2 = e^{x_1} + d + u, \\ y = x_1. \end{cases} \tag{2.31}$$

The lumped disturbances is taken as $f(x, d) = e^{x_1} + d$. The external disturbance is taken as a constant one: $d(t) = 3$ for $t \geq 6$ sec. The ESO for system (2.31) is designed as

$$\begin{cases} \dot{z}_1 = z_2 - \beta_1(z_1 - y), \\ \dot{z}_2 = z_3 - \beta_2(z_1 - y) + u, \\ \dot{z}_3 = -\beta_3(z_1 - y). \end{cases} \tag{2.32}$$

The parameters are chosen as $\beta_1 = 15$, $\beta_2 = 75$, $\beta_3 = 125$. Here, a composite control law is designed as

$$u = -4z_1 - 4z_2 - z_3.$$

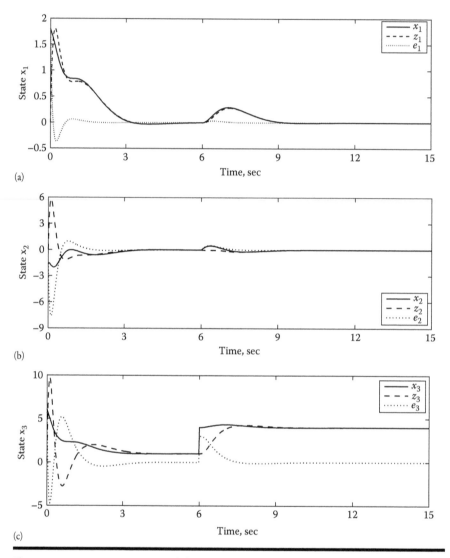

(a)

(b)

(c)

Figure 2.9   Response curves of the states and disturbance estimation for numerical example (2.31): (a) first state x_1; (b) second state x_2; (c) extended state (lumped disturbances) x_3 ($f(x, d)$).

The response curves of both states and disturbance estimation are shown in Figure 2.9. It is observed from the figure that the designed ESO can effectively estimate both states and disturbances in the presence of plant uncertainties. Note that the ESO only demands for output and input information of the system. So it is usually employed for output feedback based disturbance rejection control.

2.5 Summary

Several widely used linear-disturbance observer techniques have been introduced in this chapter, including the frequency domain disturbance observer, the time-domain disturbance observer, and the extended-state observer. The detailed design philosophy and convergence analysis also have been given in this chapter.

Chapter 3

Basic Nonlinear Disturbance Observer

3.1 Introduction

In this chapter, two kinds of practical nonlinear disturbance observers are presented. The first one is referred to as constant nonlinear disturbance observer, the estimation error of which converges to zero if the disturbance is constant one. The second kind is referred to as harmonic nonlinear disturbance observer. The disturbances in this case are not limited to constant ones anymore. The amplitudes of disturbances are not necessarily required for the observer design and analysis. However, in order to achieve accurate estimation, the frequencies of harmonic disturbances under consideration have to be known.

3.2 Nonlinear Disturbance Observer for Constant Disturbances

3.2.1 A Basic Formulation

Consider a class of affine nonlinear systems, depicted by

$$\begin{cases} \dot{x} = f(x) + g_1(x)u + g_2(x)d, \\ y = h(x), \end{cases} \qquad (3.1)$$

where $x \in R^n$, $u \in R^m$, $d \in R^l$, and $y \in R^s$ are the state, the control input, the disturbance and the output vectors, respectively. It is assumed that $f(x)$, $g_1(x)$, $g_2(x)$, and $h(x)$ are smooth functions in terms of x. The disturbances under consideration are supposed to be constant but unknown.

To estimate the unknown disturbances d, a basic nonlinear disturbance observer is suggested as [2]

$$\dot{\hat{d}} = l(x)[\dot{x} - f(x) - g_1(x)u - g_2(x)\hat{d}], \tag{3.2}$$

where $\hat{d}$ denotes the disturbance estimation vector, and $l(x)$ is the nonlinear gain function of observer.

The disturbance estimation error is defined as

$$e_d = \hat{d} - d. \tag{3.3}$$

Combining (3.3) with (3.1) and (3.2), the dynamics of disturbance estimation error are obtained, which are governed by

$$\begin{aligned}
\dot{e}_d &= \dot{\hat{d}} - \dot{d}, \\
&= l(x)[\dot{x} - f(x) - g_1(x)u - g_2(x)\hat{d}], \\
&= -l(x)g_2(x)e_d,
\end{aligned} \tag{3.4}$$

which implies that the disturbance estimation error will converge to zero as time goes to infinity if the observer gain $l(x)$ is chosen such that system (3.4) is asymptotically stable. It should be pointed out that for implementation of the above disturbance observer (3.2), the derivative of the state is required, which may need an additional sensor for measuring it.

3.2.2 An Enhanced Formulation

An enhanced nonlinear disturbance observer is then introduced to estimate the constant disturbance for system (3.1), which is depicted by [2]

$$\begin{cases} \dot{z} = -l(x)g_2(x)z - l(x)[g_2(x)p(x) + f(x) + g_1(x)u], \\ \hat{d} = z + p(x), \end{cases} \tag{3.5}$$

where $z \in R^l$ is the internal state of the nonlinear observer, and $p(x)$ is the nonlinear function to be designed. The nonlinear disturbance observer gain $l(x)$ is determined by

$$l(x) = \frac{\partial p(x)}{\partial x}. \tag{3.6}$$

The block diagram of the nonlinear disturbance observer for system (3.1) is given by Figure 3.1. Combining (3.3) with (3.1), (3.5) and (3.6), the dynamics of disturbance

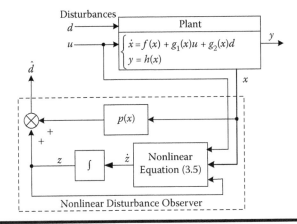

Figure 3.1 **Block diagram of the nonlinear disturbance observer for constant disturbances.**

estimation error are governed by

$$
\begin{aligned}
\dot{e}_d &= \dot{\hat{d}} - \dot{d}, \\
&= \dot{z} + \tfrac{\partial p(x)}{\partial x}\dot{x}, \\
&= -l(x)g_2(x)z - l(x)[g_2(x)p(x) + f(x) + g_1(x)u] + l(x)[f(x) \\
&\quad + g_1(x)u + g_2(x)d], \\
&= -l(x)g_2(x)\hat{d} + l(x)g_2(x)d, \\
&= -l(x)g_2(x)e_d.
\end{aligned}
\tag{3.7}
$$

It can be concluded from (3.7) that the nonlinear disturbance observer can estimate unknown constant disturbances if the observer gain $l(x)$ is chosen such that system (3.7) is asymptotically stable.

Consider the following nonlinear system with disturbances

$$
\begin{pmatrix} \dot{x}_1 \\ \dot{x}_2 \end{pmatrix} = \begin{pmatrix} x_2 \\ x_1 x_2 \end{pmatrix} + \begin{pmatrix} 0 \\ 1 + \sin^2 x_1 \end{pmatrix} u + \begin{pmatrix} 0 \\ 1 \end{pmatrix} d.
\tag{3.8}
$$

Let

$$
f(x) = \begin{pmatrix} x_2 \\ x_1 x_2 \end{pmatrix}, \quad g_1(x) = \begin{pmatrix} 0 \\ 1 + \sin^2 x_1 \end{pmatrix}, \quad g_2(x) = \begin{pmatrix} 0 \\ 1 \end{pmatrix},
$$

it can be observed that system (3.8) satisfies the form of system (3.1). Suppose that a constant disturbance is imposed on system (3.8), i.e., $d(t) = 5$ for $t \geq 6$ sec. The nonlinear disturbance observer for (3.8) is designed as (3.5). The composite

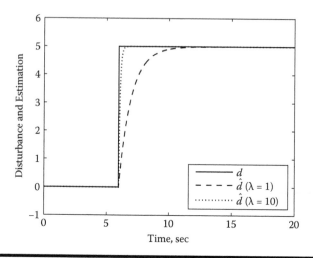

Figure 3.2 **Response curve of the disturbance estimation for numerical example (3.8) under constant disturbance for different observer parameters.**

controller is designed as

$$u = \frac{-k_1 x_1 - k_2 x_2 - x_1 x_2 - \dot{d}}{1 + \sin^2 x_1}.$$

The feedback control gains are selected as $k_1 = 10$, $k_2 = 30$. The observer gain is designed as $p(x) = \lambda x_2$. It can be calculated that $l(x)g_2(x) = \lambda$, which implies that the error dynamics of the nonlinear disturbance observer are asymptotically stable if the observer parameter λ is selected as a negative constant. The response curve of disturbance estimation for numerical example (3.8) under different observer parameters is shown by Figure 3.2.

It can be observed from Figure 3.2 that the nonlinear disturbance observer could estimate constant disturbance asymptotically. Moreover, the convergence rate of observation error dynamics can be tuned by the observer parameter. The larger the observer parameter λ, the quicker the convergence rate of observer.

3.3 Nonlinear Disturbance Observer for General Exogenous Disturbances

3.3.1 A Basic Formulation

The nonlinear system considered in this part still has the form depicted by (3.1). However, the disturbances are supposed to be harmonic ones with known frequency, but unknown amplitude and phase rather than constant ones. It is supposed that the disturbances are generated by the following exogenous system

$$\begin{cases} \dot{\xi} = A\xi, \\ d = C\xi, \end{cases} \tag{3.9}$$

where $\xi \in R^s$.

The following basic harmonic nonlinear disturbance observer can be employed to estimate the harmonic disturbances in system (3.1) and (3.9), given by [5]

$$\begin{cases} \dot{\hat{\xi}} = A\hat{\xi} + l(x)[\dot{x} - f(x) - g_1(x)u - g_2(x)\hat{d}], \\ \hat{d} = C\hat{\xi}, \end{cases} \tag{3.10}$$

where $\hat{\xi}$ is the internal state variable of observer. The observer estimation error is defined as

$$e_\xi = \hat{\xi} - \xi. \tag{3.11}$$

Combining (3.1), (3.9), (3.10), and (3.11) gives

$$\begin{aligned} \dot{e}_\xi &= \dot{\hat{\xi}} - \dot{\xi}, \\ &= A\hat{\xi} + l(x)[\dot{x} - f(x) - g_1(x)u - g_2(x)\hat{d}] - A\xi, \\ &= Ae_\xi - l(x)g_2(x)(\hat{d} - d), \\ &= [A - l(x)g_2(x)C]e_\xi. \end{aligned} \tag{3.12}$$

It is derived from (3.12) that $\hat{\xi}(t)$ approaches to $\xi(t)$ asymptotically if $l(x)$ is chosen such that (3.12) is asymptotically stable regardless of x. However, such a basic harmonic nonlinear disturbance observer still meets the problem of implementation for practical application due to the requirement of the derivatives of the states.

3.3.2 An Enhanced Formulation

Based on the original version of harmonic nonlinear disturbance observer in (3.10), an enhanced version is proposed in [5], which is depicted by

$$\begin{cases} \dot{z} = [A - l(x)g_2(x)C]z + Ap(x) - l(x)[g_2(x)Cp(x) + f(x) + g_1(x)u], \\ \hat{\xi} = z + p(x), \\ \hat{d} = C\hat{\xi}. \end{cases} \tag{3.13}$$

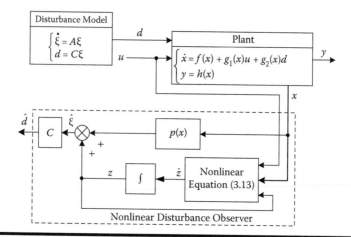

Figure 3.3 Block diagram of the nonlinear disturbance observer for harmonic disturbances.

Theorem 3.1 *Consider system (3.1) under the disturbances generated by exogenous system (3.9). The enhanced harmonic disturbance observer (3.13) can exponentially estimate the disturbances if the nonlinear observer gain $l(x)$ is selected such that*

$$\dot{e}_\xi = [A - l(x)g_2(x)C]e_\xi, \tag{3.14}$$

is globally exponentially stable regardless of x.

Proof Combining (3.1), (3.9), (3.13), and (3.11), the observation error dynamics are governed by

$$
\begin{aligned}
\dot{e}_\xi &= \dot{\hat{\xi}} - \dot{\xi} \\
&= \dot{z} + \frac{\partial p(x)}{\partial x}\dot{x} - A\xi \\
&= [A - l(x)g_2(x)C]z + Ap(x) - l(x)[g_2(x)Cp(x) + f(x) + g_1(x)u] \\
&\quad + l(x)[f(x) + g_1(x)u + g_2(x)d] - A\xi \\
&= [A - l(x)g_2(x)C](\hat{\xi} - \xi) \tag{3.15} \\
&= [A - l(x)g_2(x)C]e_\xi.
\end{aligned}
$$

It follows from (3.15) that $\hat{\xi}(t)$ approaches $\xi(t)$ exponentially if $l(x)$ is selected such that (3.14) satisfies the given condition.

Suppose that the relative degree from the disturbance to the output, r, is uniformly well-defined. This implies that $L_{g2}L_f^{r-1}h(x) \neq 0$ for all x, where L denotes Lie derivatives. Without loss of generality, suppose that $L_{g2}L_f^{r-1}h(x) > 0$ in the following analysis, which implies that $L_{g2} L_f^{r-1} h(x)$ can be divided as

$$L_{g2} L_f^{r-1} h(x) = \alpha_0 + \alpha_1(x), \qquad (3.16)$$

where $\alpha_0 > 0$ is a constant that can be chosen as the minimum of function $L_{g2} L_f^{r-1} h(x)$ over all x, and $\alpha_1(x) > 0$ for all x. The nonlinear function $p(x)$ is designed as

$$p(x) = K L_f^{r-1} h(x), \qquad (3.17)$$

where $K = [k_1, \cdots, k_m]^T$ are gains to be determined. It follows from (3.6) that

$$l(x) = \frac{\partial p(x)}{\partial x} = K \frac{L_f^{r-1} h(x)}{\partial x}. \qquad (3.18)$$

Substituting the above observer gain (3.18) and (3.16) into the observation error dynamics (3.14) gives

$$
\begin{aligned}
\dot{e}_\xi &= \left[A - K \frac{L_f^{r-1} h(x)}{\partial x} g_2(x) C \right] e_\xi, \\
&= [A - K L_{g2} L_f^{r-1} h(x) C] e_\xi, \qquad (3.19) \\
&= [A - K(\alpha_0 + \alpha_1(x)) C] e_\xi.
\end{aligned}
$$

Theorem 3.2 *The estimation $\hat{d}$ yielded by harmonic nonlinear disturbance observer (3.13) converges to the disturbance d globally exponentially if there exists a gain K such that the transfer function*

$$H(s) = C(s I - \bar{A})^{-1} K, \qquad (3.20)$$

is asymptotically stable and strictly positive real where

$$\bar{A} = (A - K \alpha_0 C). \qquad (3.21)$$

Proof According to strictly positive real Lemma, transfer function (3.20) being stable and positive real implies that there exists a positive definite matrix P such that

$$\bar{A}^T P + P \bar{A} < 0, \qquad (3.22)$$

and

$$P K = C^T. \qquad (3.23)$$

A Lyapunov candidate function for observation error dynamics (3.12) is defined as

$$V(e_\xi) = e_\xi^T P e_\xi. \qquad (3.24)$$

Taking derivative of the Lyapunov function with respect to time along the trajectory of observer error dynamics (3.12) gives

$$
\begin{aligned}
\dot{V}(e_\xi) &= 2e_\xi^T P[A - K(\alpha_0 + \alpha_1(x))C]e_\xi \\
&= e_\xi^T(\bar{A}^T P + P\bar{A})e_\xi - 2e_\xi^T PKCe_\xi\alpha_1(x) \\
&< -\delta e_\xi^T e_\xi - 2e_\xi^T C^T Ce_\xi\alpha_1(x),
\end{aligned}
\tag{3.25}
$$

where δ is a small positive scalar depending on (3.22). Since the relative degree from disturbance to output is uniformly well defined, it follows from (3.16) that $\alpha_1(x) > 0$ regardless of x. Noting that $e_\xi^T C^T Ce_\xi \geq 0$, it follows from (3.25) that

$$
\dot{V}(e_\xi) < -\delta e_\xi^T e_\xi,
\tag{3.26}
$$

for any x and e, which implies that estimation $\hat{d}$ yielded by nonlinear disturbance observer (3.13) approaches to disturbance d globally exponentially.

Theorem 3.2 states that the convergence of disturbance observer regardless of x can be guaranteed by determining the gain K such that transfer function (3.20) is asymptotically stable and strictly positive real. According to the proof, this can be performed by finding a suitable P and K such that conditions (3.22) and (3.23) are satisfied. Combining (3.21) with (3.22) and (3.23), it can be shown that conditions in (3.22) and (3.23) are satisfied if

$$
A^T P + PA - 2\alpha_0 C^T C < 0,
\tag{3.27}
$$

which is easy to calculate by using linear matrix inequalities (LMIs) packages. After P is determined, the gain can be calculated by

$$
K = P^{-1}C^T.
\tag{3.28}
$$

The nonlinear system expressed by (3.8) is employed here for simulation studies, where disturbance d is governed by an exogenous system that has the form of (3.9), with

$$
A = \begin{bmatrix} 0 & 2 \\ -2 & 0 \end{bmatrix}, \quad C = \begin{bmatrix} 1 & 0 \end{bmatrix}.
$$

The nonlinear disturbance observer for system (3.8) with disturbance (3.9) is designed as (3.13). The composite controller is still designed as

$$
u = \frac{-k_1 x_1 - k_2 x_2 - x_1 x_2 - \hat{d}}{1 + \sin^2 x_1}.
$$

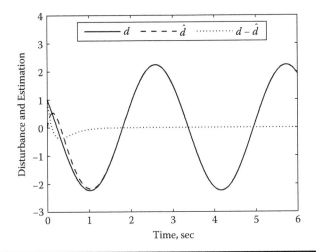

Figure 3.4   Response curve of the disturbance estimation for numerical example (3.8) under general exogenous disturbance (3.9).

The feedback control gains are selected as $k_1 = 10$, $k_2 = 30$. The observer gain vector is designed as $p(x) = [\lambda_1 x_2, \lambda_2 x_2]^T$. It can be calculated that

$$A_\xi = A - l(x)g_2(x)C = \begin{bmatrix} -\lambda_1 & 2 \\ -2 - \lambda_2 & 0 \end{bmatrix},$$

which implies that the error dynamics of nonlinear disturbance observer are asymptotically stable if the observer parameters λ_1 and λ_2 are selected such that A_ξ is a Hurwitz matrix. In this numerical simulation example, the observer parameters are selected as $\lambda_1 = 10$ and $\lambda_2 = 10$, respectively. It can be calculated that the eigenvalues of matrix A_ξ are -4 and -6, respectively, which implies that A_ξ is a Hurwitz matrix, and the observation errors are asymptotically stable. The response curve of disturbance estimation for numerical example (3.8) with exogenous disturbance (3.9) is shown by Figure 3.4.

It can be observed from Figure 3.4 that the nonlinear disturbance observer could asymptotically estimate the general exogenous disturbance governed by (3.9). The convergence rate of observation error dynamics can be tuned by the observer parameter.

3.4 Summary

Two kinds of traditional nonlinear disturbance observer techniques have been discussed in this chapter. Firstly, a constant nonlinear disturbance observer, which has been widely employed in industrial applications, has been introduced first. It has been proved that the disturbance estimation error will converge to zero if the disturbance is constant. Secondly, a harmonic nonlinear disturbance observer has also been discussed in this chapter. The disturbances under consideration are supposed to be harmonic with known frequencies but unknown amplitudes. In this case, the disturbance estimation error tends to zero regardless of the time varying property of disturbances.

Chapter 4

Advanced Nonlinear Disturbance Observer

4.1 Introduction

Followed by the basic nonlinear disturbance observer introduced in the last chapter, three advanced nonlinear disturbance observer techniques are illustrated in this chapter. The first one concerns a high-order time varying disturbance case, which implies that the disturbance can be expressed in time series expansion. The second one focuses on the output design of the nonlinear disturbance observer, which implies that only the output but not the state information is employed. The method is inspired by the ideas of extended state observer and high-gain observer methods. Finally, the finite-time disturbance observer technique is discussed. This observer employs the high-order sliding mode differentiator techniques. The estimation error of observer will converge to zero in finite time, which shows a much faster convergence rate than other types of disturbance observers.

4.2 High-Order Disturbance Observer

The result of this section is a summarization of the high-order disturbance observer proposed in [11]. The readers can refer to [11] for more detailed illustration.

4.2.1 Constant Disturbance Case

Consider a class of nonlinear systems, depicted by

$$\dot{x} = f(x, u; t) + Fd(t), \tag{4.1}$$

53

where $x \in R^n$, $u \in R^m$, and $d \in R^r$ are the state, the control input, and the disturbance vectors, respectively. $f(x, u; t)$ and the matrix F with $rank(F) = r$ are known. The disturbances $d(t)$ are supposed to be constant ones here. It is assumed that the state variables x are measurable and their initial values $x(0) = x_0$ are known. A reduced order system of (4.1) is expressed as

$$F^+ \dot{x} = F^+ f(x, u; t) + d(t), \tag{4.2}$$

where F^+ denotes the Moore-Penrose pseudo-inverse of matrix F.

A constant disturbance observer for system (4.1) has been proposed in [11], expressed as

$$\begin{cases} \dot{z} = F^+ f(x, u; t) + \Gamma_0(F^+ x - z), \\ \hat{d} = \Gamma_0(F^+ x - z), \end{cases} \tag{4.3}$$

where $\hat{d}$ denotes the disturbance estimation vector, and $\Gamma_0 = diag\{\gamma_{01}, \dots, \gamma_{0r}\}$, $\gamma_{0i} > 0$ $(i = 1, \dots, r)$ the observer gain to be designed.

The disturbance estimation error is defined as

$$e_d = \hat{d} - d. \tag{4.4}$$

Combining (4.4) with (4.2) and (4.3), the dynamics of disturbance estimation error are derived, given by

$$\dot{e}_d = -\Gamma_0 e_d, \tag{4.5}$$

which implies that the disturbance estimation error will converge to zero as time goes to infinity if the observer gain $-\Gamma_0$ is chosen as a Hurwitz matrix.

Consider a simple first-order dynamic system

$$\dot{x} = -2x + d(t), \tag{4.6}$$

where $d(t)$ is the external disturbance.

Suppose that the system is subject to a square wave disturbance, a constant disturbance observer is constructed as

$$\begin{cases} \dot{z} = -2x + r_0(x - z), \\ \hat{d} = r_0(x - z), \end{cases} \tag{4.7}$$

The response curves of disturbance estimation under different observer parameters are shown in Figure 4.1. It is shown by the figure that the performance of constant DO is determined by observer parameter r_0. It can be observed that as r_0 increases, the DO results in a faster convergence rate for disturbance estimation.

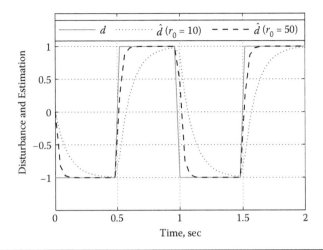

Figure 4.1 Response curves of disturbance estimation for system (4.6) with a square wave disturbance under different observer parameters.

4.2.2 Ramp Disturbance Case

Suppose that disturbances $d(t)$ in (4.1) are ramp ones, which are expressed as

$$d(t) = d_0 + d_1 t, \tag{4.8}$$

where d_0 and d_1 are constant but unknown. A ramp disturbance observer proposed in [11] is given by

$$
\begin{cases}
\dot{z} = F^+ f(x, u; t) + \Gamma_0 (F^+ x - z) + \Gamma_1 \displaystyle\int_0^t (F^+ x - z) d\tau, \\[4mm]
\hat{d} = \Gamma_0 (F^+ x - z) + \Gamma_1 \displaystyle\int_0^t (F^+ x - z) d\tau,
\end{cases}
\tag{4.9}
$$

where $\Gamma_1 = diag\{\gamma_{11}, \dots, \gamma_{1r}\}$, and γ_{ij}'s are chosen such that the polynomials $p_j(s) = s^2 + \gamma_{0j}s + \gamma_{1j}$ $(j = 1, \dots, r)$ are Hurwitz stable.

Combining (4.2) with (4.4), (4.8) and (4.9), the dynamics of disturbance estimation error are derived as follows:

$$
\begin{aligned}
\dot{e}_d &= \dot{\hat{d}} - \dot{d}, \\
&= \Gamma_0 (F^+ \dot{x} - \dot{z}) + \Gamma_1 (F^+ x - z) - \dot{d}, \\
&= -\Gamma_0 e_d + \Gamma_1 (F^+ x - z) - \dot{d}.
\end{aligned}
\tag{4.10}
$$

Taking derivative of both sides of (4.10) in terms of time along system trajectory (4.2) and also observer dynamics (4.9) gives

$$\ddot{e}_d = -\Gamma_0 \dot{e}_d + \Gamma_1(F^+ \dot{x} - \dot{z}) - \ddot{d},$$
$$= -\Gamma_0 \dot{e}_d - \Gamma_1 e_d - \ddot{d}. \tag{4.11}$$

Since matrices Γ_0 and Γ_1 are chosen as diagonal ones, the observer error dynamics are decoupled. Consider the case that the disturbances are ramp ones, that is, $\ddot{d}(t) = 0$, it is obtained from (4.11) that the error dynamics of jth observer are governed by

$$\ddot{e}_{dj} + \gamma_{0j} \dot{e}_{dj} + \gamma_{1j} e_{dj} = 0, \tag{4.12}$$

which implies that the convergent dynamics of observer error system (4.12) can be separately tuned by assigning the poles of the polynomial equations, $s^2 + \gamma_{0j} s + \gamma_{1j} = 0$ for $j = 1, \ldots, r$.

System (4.6) that subjects to a sawtooth disturbance (representative of the ramp disturbance) is taken as numerical example for simulation studies. Actually, the disturbance is represented as follows

$$d(t) = 1 - t, \text{ for } 0 \le t \le 1; \ d(t+1) = d(t), \text{ for } t > 1.$$

The ramp disturbance observer is constructed as

$$\begin{cases} \dot{z} = -2x + r_0(x - z) + r_1 \int_0^t (x - z)dt, \\ \hat{d} = r_0(x - z) + r_1 \int_0^t (x - z)dt. \end{cases} \tag{4.13}$$

The response curves of disturbance estimation under different observer parameters are shown in Figure 4.2. As shown by this figure, the performance of the ramp DO is determined by the observer parameters r_0 and r_1. Note that observer parameters $r_0 = 20$, $r_1 = 100$ imply that the eigenvalues of observer error dynamics are $s_1 = s_2 = -10$, while parameters $r_0 = 40$, $r_1 = 200$ imply that the eigenvalues of observer error dynamics are $s_1 = s_2 = -20$. It can be observed from Figure 4.2 that a quicker convergence rate is obtained if the eigenvalues of observer error dynamics are assigned farer from the imaginary axis.

If we set $r_1 = 0$, ramp disturbance observer (4.13) reduces to constant disturbance observer (4.7). For the purpose of comparison, two sets of observer parameters r_0, r_1 are considered: $r_0 = 20$, $r_1 = 0$ for the constant disturbance observer and $r_0 = 40$, $r_1 = 400$ for the ramp disturbance observer, respectively. The response curves of disturbance estimation for system (4.6) with a sawtooth disturbance by ramp and constant DOs are shown by Figure 4.3. It is observed that the ramp DO could asymptotically estimate the disturbance, while the constant disturbance observer results in a steady state error in disturbance estimation.

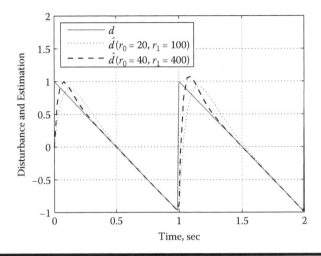

Figure 4.2 Response curves of disturbance estimation for system (4.6) with a sawtooth disturbance under different observer parameters.

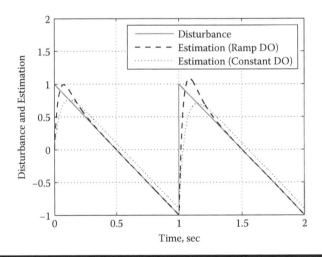

Figure 4.3 Response curves of disturbance estimation for system (4.6) with a sawtooth disturbance by ramp and constant DOs.

4.2.3 High-Order Disturbance Case

Disturbances $d(t)$ in (4.1) are assumed to be high-order ones, which are depicted by

$$d(t) = d_0 + d_1 t + \cdots + d_q t^q, \tag{4.14}$$

where $d_0, d_1, \ldots, d_q$ are constant but unknown. A high-order disturbance observer for estimating the disturbances in (4.14) is proposed in [11], given by

$$
\begin{cases}
\dot{z} = F^+ f(x, u; t) + \Gamma_0 g_0(t) + \cdots + \Gamma_q g_q(t), \\
\hat{d} = \Gamma_0 g_0(t) + \cdots + \Gamma_q g_q(t),
\end{cases}
\tag{4.15}
$$

with

$$
g_k(t) =
\begin{cases}
F^+ x - z, & (k = 0), \\
\int_0^t g_{k-1}(\tau)d\tau, & (k \geq 1),
\end{cases}
\tag{4.16}
$$

for $k \in [0, q]$, where $\Gamma_k = diag\{\gamma_{k1}, \ldots, \gamma_{kr}\}$, $(k = 0, \ldots, q)$, and γ_{ij}'s are chosen such that the polynomials

$$p_j(s) = s^{q+1} + \gamma_{0j} s^q + \cdots + \gamma_{(q-1)j} s + \gamma_{qj} \tag{4.17}$$

for $j = 1, \ldots, r$ are Hurwitz stable.
Combining (4.2) with (4.4), (4.14)–(4.16), the dynamics of disturbance estimation error are derived as follows:

$$
\begin{aligned}
\dot{e}_d &= \hat{d} - \dot{d}, \\
&= \Gamma_0 g_0(t) + \Gamma_1 g_1(t) + \cdots + \Gamma_q g_q(t) - \dot{d}, \\
&= -\Gamma_0 e_d + \Gamma_1 g_0(t) + \cdots + \Gamma_q g_{q-1}(t) - \dot{d}.
\end{aligned}
\tag{4.18}
$$

Taking derivative of (4.18) in terms of time along system trajectory (4.2) and also the observer dynamics (4.15) and (4.16) gives

$$
\begin{aligned}
\ddot{e}_d &= -\Gamma_0 \dot{e}_d + \Gamma_1 \dot{g}_0(t) + \Gamma_2 g_1(t) + \cdots + \Gamma_q g_{q-1}(t) - \ddot{d}, \\
&= -\Gamma_0 \dot{e}_d - \Gamma_1 e_d + \Gamma_2 g_0(t) + \cdots + \Gamma_q g_{q-2}(t) - \ddot{d}.
\end{aligned}
\tag{4.19}
$$

Keep calculating the derivatives of observer error dynamics (4.19) until the appearance of $d^{[q]}$, gives

$$e_d^{[q]} = -\Gamma_0 e_d^{[q-1]} - \Gamma_1 e_d^{[q-2]} - \cdots - \Gamma_q e_d - d^{[q]}. \tag{4.20}$$

Since matrices Γ_k $(k = 0, \ldots, q)$ are chosen as diagonal ones, the dynamics of observer error are also decoupled here. Consider the case that the disturbances satisfy the formation in (4.14), that is, $d^{[q]}(t) = 0$, it is obtained from (4.20) that the error dynamics of jth observer are governed by

$$e_{dj}^{[q]} + \gamma_{0j}e_{dj}^{[q-1]} + \gamma_{1j}e_{dj}^{[q-2]} + \cdots + \gamma_{qj}e_{dj} = 0. \tag{4.21}$$

which implies that the convergent dynamics of observer error system (4.21) can be separately tuned by assigning the poles of the polynomial equations, $p_j(s) = 0$ for $j = 1, \ldots, r$.

System (4.6) that subjects to a high-order disturbance is taken as a numerical example for this simulation studies. Actually, the disturbance is represented as follows

$$d(t) = t - t^2, \text{ for } 0 \le t \le 1; \ d(t+1) = d(t), \text{ for } t > 1.$$

In this case, a high-order disturbance observer is constructed as

$$\begin{cases} \dot{z} = -2x + r_0(x - z) + r_1 \int_0^t (x - z)dt + r_2 \int_0^t \int_0^\tau (x - z)d\tau\, dt, \\ \hat{d} = r_0(x - z) + r_1 \int_0^t (x - z)dt + r_2 \int_0^t \int_0^\tau (x - z)d\tau\, dt \end{cases} \tag{4.22}$$

The response curves of disturbance estimation under different observer parameters are shown in Figure 4.4. As shown by this figure, the performance of the ramp DO is determined by the observer parameters r_i ($i = 0, 1, 2$). Note that observer parameters $r_0 = 30$, $r_1 = 300$, $r_2 = 1000$ imply that the eigenvalues of observer error dynamics are $s_1 = s_2 = s_3 = -10$, while parameters $r_0 = 60$, $r_1 = 1200$, $r_2 = 8000$ imply that the eigenvalues of observer error dynamics are $s_1 = s_2 = s_3 = -20$. It can be observed from Figure 4.4 that a quicker convergence rate is obtained if the eigenvalues of observer error dynamics are placed farer away from the imaginary axis.

If we set $r_2 = 0$, high-order DO (4.22) reduces to ramp DO (4.13). For the purpose of comparison studies, both constant and ramp DOs are employed here to estimate such a high-order disturbance. For the constant and ramp DOs, the observer parameters are selected as $r_0 = 10$ and $r_0 = 20$, $r_1 = 100$, respectively. The parameters of the high-order DO are designed as $r_0 = 60$, $r_1 = 1200$, $r_2 = 8000$. The response curves of disturbance estimation for system (4.6) with a high-order disturbance by these three DOs are shown by Figure 4.5. It is observed that the high-order DO could asymptotically estimate the disturbance, while both the constant and the ramp DOs lead to a large steady state error in estimation for such a high-order disturbance.

4.3 Extended High-Gain State Observer

The results of this section mainly refer to the extended high-gain state observer based control proposed in [7]. Consider a class of single-input single-output nonlinear

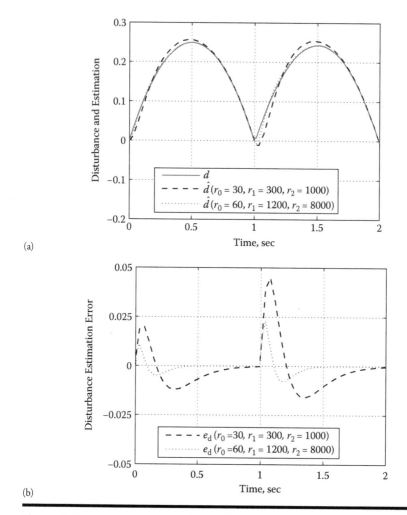

(a)

(b)

Figure 4.4 **Response curves of disturbance estimation for system (4.6) under a high-order disturbance for different observer parameters: (a) disturbance and estimation; (b) disturbance estimation error.**

systems in the following normal form [106]:

$$\begin{cases} \dot{x} = Ax + B\left[b(x, z) + a(x, z)u + d\right], \\ \dot{z} = f_0(x, z, d), \\ y = Cx, \end{cases} \quad (4.23)$$

where $x \in R^n$ and $z \in R^m$ are the state variables, $u \in R$ is the control input, $y \in R$ is the measured output, and $d \in R$ is a disturbance. Nonlinear functions $a(\cdot)$ and $b(\cdot)$ are assumed to be continuously differentiable with locally Lipschitz derivatives, $a(\cdot) \geq a_0$ with a known $\alpha_0 > 0$, and $f_0(\cdot)$ is locally Lipschitz. The disturbance

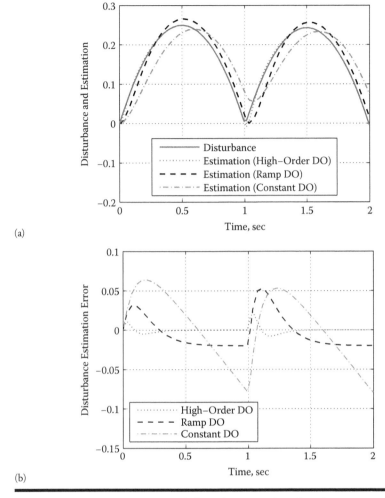

(a)

(b)

Figure 4.5 Response curves of disturbance estimation for system (4.6) with a high-order disturbance by three DOs: (a) disturbance and estimation; (b) disturbance estimation error.

$d(t)$ is supposed to belong to a known compact set $W \subset R$ and $\dot{d}(t)$ is bounded. $A \in R^{n \times n}$, $B \in R^{n \times 1}$, and $C \in R^{1 \times n}$ are system matrices, depicted by

$$
A = \begin{bmatrix} 0 & 1 & 0 & \cdots & 0 \\ 0 & 0 & 1 & \cdots & 0 \\ \vdots & \vdots & \ddots & \ddots & \vdots \\ 0 & 0 & \cdots & 0 & 1 \\ 0 & 0 & \cdots & \cdots & 0 \end{bmatrix}, \quad B = \begin{bmatrix} 0 \\ 0 \\ \vdots \\ 0 \\ 1 \end{bmatrix}, \quad C = \begin{bmatrix} 1 \\ 0 \\ \vdots \\ 0 \\ 0 \end{bmatrix}^{T} .
$$

It is also supposed that system (4.23) is of minimum phase, which implies that the zero dynamics of system (4.23) are stable. This implies that there exists a radially unbounded positive definite function $V_0(z)$ such that for all $x \in R^n$, $z \in R^m$ and $d \in W$

$$\frac{\partial V_0(z)}{\partial z} f_0(x, z, d) \leq 0, \text{ for } \|z\| \geq \chi(x, d),$$

where $\chi(x, d)$ is a nonnegative continuous function.

In the case of known (x, z, d), the following feedback control law

$$u = \frac{-b(x, z) - d + v}{a(x, z)}, \tag{4.24}$$

can linearize the original system to

$$\dot{x} = Ax + Bv, \quad y = Cx, \tag{4.25}$$

where the auxiliary control v is generally designed as a feedback function in terms of x to stabilize the original system and ensure the closed-loop system achieve a satisfactory transient response. Let $v = \phi(x)$ be a twice differentiable state feedback control law such that the closed-loop system

$$\dot{x} = Ax + B\phi(x), \tag{4.26}$$

is locally exponentially stable and globally asymptotically stable. Let $V_s(x)$ be a radially unbounded Lyapunov function such that

$$\frac{\partial V_s(x)}{\partial x} [Ax + B\phi(x)] \leq -\Psi(x), \quad \forall x,$$

for some positive definite function $\Psi(x)$.

The control object of this design is to track the following target system

$$\dot{x}^\star = Ax^\star + B\phi(x^\star), \quad x^\star(0) = x(0). \tag{4.27}$$

To estimate the disturbance in (4.23), an additional integrator $x_{n+1} = d$ is augmented, and system (4.23) is rewritten as

$$\begin{cases} \dot{x} &= Ax + B[b(x, z) + a(x, z)u + x_{n+1}], \\ \dot{x}_{n+1} &= \dot{d}(t), \\ \dot{z} &= f_0(x, z, d), \\ y &= Cx. \end{cases} \tag{4.28}$$

An extended high gain observer for system (4.23) is then designed as [7]

$$
\begin{cases}
\dot{\hat{x}} &= A\hat{x} + B\left[\hat{b}(\hat{x}) + \hat{a}(\hat{x})u + \hat{x}_{n+1}\right] + H(\varepsilon)(y - C\hat{x}), \\
\dot{\hat{x}}_{n+1} &= \frac{\alpha_{n+1}}{\varepsilon^{n+1}}(y - C\hat{x}),
\end{cases}
\tag{4.29}
$$

where $\hat{a}(\cdot) \geq a_0 > 0$ and $\hat{b}(\cdot)$ are twice continuously differentiable and globally bounded functions, respectively. $H(\varepsilon)$ is designed as

$$
H(\varepsilon) = \left[\frac{\alpha_1}{\varepsilon}, \cdots, \frac{\alpha_n}{\varepsilon^n}\right]^T,
$$

where $\varepsilon > 0$ is a small constant, and $\alpha_1, \ldots, \alpha_n, \alpha_{n+1}$ are chosen such that the polynomial

$$
s^{n+1} + \alpha_1 s^n + \cdots + \alpha_n s + \alpha_{n+1}
$$

is Hurwitz.

The corresponding output-feedback based control law is designed as

$$
u = \frac{-\hat{x}_{n+1} - \hat{b}(\hat{x}) + \phi(\hat{x})}{\hat{a}(\hat{x})} \triangleq \psi(\hat{x}, \hat{x}_{n+1}).
\tag{4.30}
$$

However, from the high-gain observer theory, the control outside a compact set of interest is generally saturated to overcome the peaking phenomenon in the observer's transient response. It can be shown that there exist a constant $c_0 \geq 0$ and a class $\mathcal{K}_\infty$ function $\alpha(\cdot)$ such that for any positive constant c, $\{V_0(z) \leq c_0 + \alpha(c)\}$ is a positively invariant set of $\dot{z} = f_0(x, z, d)$ for all $x(t) \in \{V_s(x) \leq c\}$ and $d(t) \in W$. Define a compact set Ω_c by

$$
\Omega_c = \{V_s(x) \leq c\} \times \{V_0(z) \leq c_0 + \alpha(c)\}.
$$

The final saturated control law is designed as

$$
u = M\,\mathrm{sat}\left(\frac{\psi(\hat{x}, \hat{x}_{n+1})}{M}\right),
\tag{4.31}
$$

where $\mathrm{sat}(\cdot)$ is a standard saturation function, defined by

$$
\mathrm{sat}(y) = \begin{cases}
y & \text{when } |y| \leq 1, \\
1 & \text{when } y > 1, \\
-1 & \text{when } y < -1,
\end{cases}
$$

and

$$M > \max_{(x, z) \in \Omega_c, d \in W} \left| \frac{-b(x, z) - d + \phi(x)}{a(x, z)} \right|.$$

The closed-loop system under the output feedback control law (4.31) recovers the performance of target system (4.27). Let

$$k_a = \max_{(x, z) \in \Omega_c, d \in W} \left| \frac{a(x, z) + d - \hat{a}(x)}{\hat{a}(x)} \right|,$$

$$G(s) = \frac{\alpha_{n+1}}{s^{n+1} + \alpha_1 s^n + \cdots + \alpha_{n+1}},$$

and $\|G\|_\infty = \sup_\omega |G(j\omega)|$. The main results are stated as follows [7].

Theorem 4.1 *Consider the closed-loop system formed of the plant (4.23), the observer (4.29) and the controller (4.31). Suppose that the parameters are chosen such that*

$$k_a < \frac{1}{\|G\|_\infty},$$

the initial states of the observer belong to a compact subset of R^{n+1}, and the initial states of plant belong to a compact set in the interior of Ω_c. Then, there exists $\bar{\varepsilon} > 0$ such that for $\varepsilon \in (0, \bar{\varepsilon})$:

- *All trajectories are bounded;*
- *$\|x(t) - x^*(t)\| \to 0$ as $\varepsilon \to 0$, uniformly in t, for $t \geq 0$;*
- *$\|x(t)\|$ is uniformly ultimately bounded by $\delta(\varepsilon)$, where $\delta(\varepsilon) \to 0$ as $\varepsilon \to 0$.*

The above theorem shows that the regulation error is ultimately bounded by $\delta(\varepsilon)$, which can be made arbitrarily small by choosing sufficiently small ε. This implies that the extended high-gain observer based control method can achieve transient performance recovery by setting a sufficiently small parameter ε. The proof of this theorem is involved with singular perturbation theory, circle criterion, and Kalman–Yakubovich–Popov lemma. The proof process is omitted here and the readers can refer to [7] for the details.

Consider the following numerical example for simulation verification

$$\begin{cases} \dot{x}_1 &= x_2, \\ \dot{x}_2 &= \sin x_1 + e^{x_1} + u + d, \\ y &= x_1 \end{cases} \tag{4.32}$$

where $d(t)$ is the external disturbance.

An extended high-gain state observer for system (4.32) is designed as

$$\begin{cases} \dot{z}_1 &= z_2 + 8(y - z_1)/\varepsilon, \\ \dot{z}_2 &= sin(y) + e^y + u + z_3 + 12(y - z_1)/\varepsilon^2, \\ \dot{z}_3 &= 6(y - z_1)/\varepsilon^3, \end{cases} \qquad (4.33)$$

where the parameter is selected as $\varepsilon = 0.1$.

A composite controller for this system is designed as

$$u = M\,\text{sat}\left(\frac{-\sin z_1 - e^{z_1} - 4z_1 - 4z_2 - z_3}{M}\right),$$

where the parameter M is chosen as 50.

The external disturbance is taken as a constant one: $d = 5$ for $t \geq 5$ sec. The response curves of both the states and the disturbance estimation are shown in Figure 4.6.

It is observed from Figure 4.6 that the designed extended high-gain state observer can effectively estimate both the states and the extended state in the presence of external disturbance.

4.4 Finite-Time Disturbance Observer

The contents of this section are mainly referred to [8]. The nonlinear system considered in this section still has the form of (4.1). However, the matrix F is supposed to satisfy $rank(F) = r \leq n$. Disturbance $d(t)$ is supposed to be lth differentiable, and $d^{[l]}(t)$ has a Lipshitz constant L. Let $w(t) = Fd(t)$, system (4.1) is rewritten as

$$\dot{x} = f(x, u; t) + w(t). \qquad (4.34)$$

A finite-time disturbance observer for estimating disturbance $w(t)$ in system (4.34) has been proposed in [8], given by

$$\begin{cases} \dot{z}_0 = v_0 + f(x, u, t), \ v_0 = -\lambda_0 L^{1/(l+1)}|z_0 - x|^{l/(l+1)}\text{sign}(z_0 - x) + z_1, \\ \dot{z}_1 = v_1, \ v_1 = -\lambda_1 L^{1/l}|z_1 - v_0|^{l-1/l}\text{sign}(z_1 - v_0) + z_2, \\ \qquad \vdots \\ \dot{z}_{l-1} = v_{l-1}, \ v_{l-1} = -\lambda_{l-1} L^{1/2}|z_{l-1} - v_{l-2}|^{1/2}\text{sign}(z_{l-1} - v_{l-2}) + z_l, \\ \dot{z}_l = v_l, \ v_l = -\lambda_l L\text{sign}(z_l - v_{l-1}), \\ \hat{x} = z_0, \ \hat{w} = z_1, \dots, \hat{w}^{[l-1]} = z_l, \end{cases} \qquad (4.35)$$

where $\lambda_i > 0$ $(i = 0, 1, \dots, l)$ are the observer coefficients to be designed, $\hat{x}$, $\hat{w}$, $\dots$, $\hat{w}^{[r-1]}$ are the estimates of x, w, $\dots$, $w^{[r-1]}$, respectively.

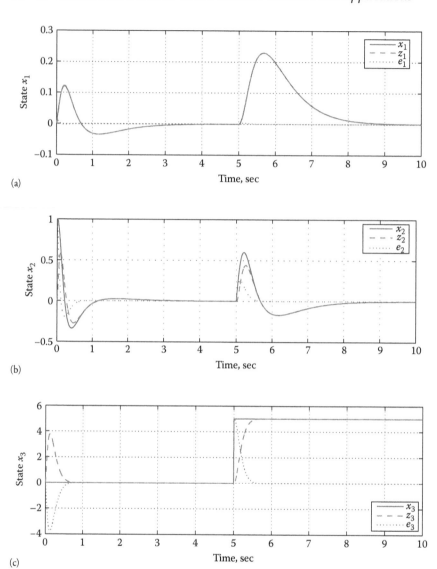

Figure 4.6 **Response curves of the states and the disturbance estimation for numerical example (4.32): (a) first state x_1; (b) second state x_2; (c) extended state (disturbances) $x_3 = d$.**

Define the disturbance estimation errors as

$$e_0 = z_0 - x, \ e_1 = z_1 - w, \ \ldots, \ e_{l-1} = z_{l-1} - w^{[l-2]}, \ e_l = z_l - w^{[l-1]}. \quad (4.36)$$

Combining (4.34) with (4.35) and (4.36), the dynamics of observer estimation error are obtained, which are governed by

$$\begin{cases} \dot{e}_0 = -\lambda_0 L^{1/(l+1)} |e_0|^{l/(l+1)} \mathrm{sign}(e_0) + e_1, \\ \dot{e}_1 = -\lambda_1 L^{1/l} |e_1 - \dot{e}_0|^{l-1/l} \mathrm{sign}(e_1 - \dot{e}_0) + e_2, \\ \quad \vdots \\ \dot{e}_{l-1} = -\lambda_{l-1} L^{1/2} |\dot{e}_{l-1} - \dot{e}_{l-2}|^{1/2} \mathrm{sign}(\dot{e}_{l-1} - \dot{e}_{l-2}) + e_l, \\ \dot{e}_l = -\lambda_l L \mathrm{sign}(\dot{e}_l - \dot{e}_{l-1}) + [-L, L]. \end{cases} \quad (4.37)$$

It follows from [8] that observer error system (4.37) is finite-time stable, that is, there exists a time constant $t_f > t_0$ such that $e_i(t) = 0$ ($i = 0, 1, \ldots, l$) (or equivalently $\hat{x}(t) = x(t)$, $\hat{w}(t) = w(t)$, $\ldots$, $\hat{w}^{[l-1]}(t) = w^{[l-1]}(t)$) for $t \geq t_f$.

The disturbance estimation of original system (4.1) is defined as

$$\hat{d} = \left(F^T F\right)^{-1} F^T \hat{w}, \quad (4.38)$$

where $F^T F$ is definitely invertible due to the rank condition that $\mathrm{rank}(F) = r \leq n$.

Then, the disturbance estimation error of original system (4.1) is derived as

$$e_d = \hat{d} - d = \left(F^T F\right)^{-1} F^T \hat{w} - \left(F^T F\right)^{-1} F^T w = \left(F^T F\right)^{-1} F^T e_1. \quad (4.39)$$

It is concluded from the finite-time stability of observer error system (4.37) that the disturbance estimation error $e_d(t)$ converges to zero in finite time t_f.

Consider the following numerical simulation example

$$\begin{cases} \dot{x}_1 = x_2, \\ \dot{x}_2 = -4x_1 - 4x_2 + d, \end{cases} \quad (4.40)$$

where $d(t)$ is the external disturbance.

The disturbance under consideration is taken as $d = 0.5 + 0.5 \sin t$ for system (4.40). A finite-time disturbance observer for (4.40) is designed as

$$\begin{cases} \dot{z}_0 = v_0 + x_2, \ \dot{z}_1 = v_1, \ \dot{z}_2 = v_2, \ \dot{z}_3 = v_3, \\ v_0 = -\lambda_0 L^{1/4} |z_0 - x_1|^{3/4} \mathrm{sgn}(z_0 - x_1) + z_1, \\ v_1 = -\lambda_1 L^{1/3} |z_1 - v_0|^{2/3} \mathrm{sgn}(z_1 - v_0) + z_2, \\ v_2 = -\lambda_2 L^{1/2} |z_2 - v_1|^{1/2} \mathrm{sgn}(z_2 - v_1) + z_3, \\ v_3 = -\lambda_3 L \mathrm{sgn}(z_3 - v_2), \\ \hat{x}_2 = z_0, \ \hat{d} = z_1, \ \dot{\hat{d}} = z_2, \ \ddot{\hat{d}} = z_3, \end{cases} \quad (4.41)$$

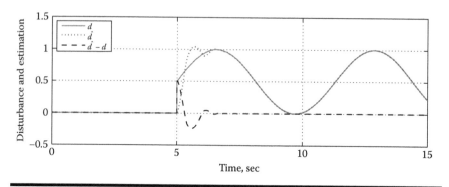

Figure 4.7 **Response curves of disturbance estimation for numerical example (4.40) by finite-time DO.**

where the parameters of observer (4.41) are designed as

$$\lambda_0 = 8, \lambda_1 = 3, \lambda_2 = 1.5, \lambda_3 = 1.1, L = 5.$$

The response curves of disturbance estimation are shown in Figure 4.7.

It is observed from Figure 4.7 that the designed finite-time disturbance observer can effectively estimate the disturbance.

4.5 Summary

Three kinds of advanced nonlinear disturbance observers, which can improve the disturbance estimation properties from different aspects, have been discussed in this chapter. First, a high-order disturbance observer has been introduced, which enlarges the disturbance types to be handled. Second, the output-based disturbance observer design problem has been discussed, where an extended high-gain state observer has been introduced. Finally, a disturbance observer with finite-time convergence rate has been introduced.

DISTURBANCE OBSERVER-BASED CONTROL DESIGN

Chapter 5

Disturbance Observer-Based Control for Nonlinear Systems

5.1 Introduction

This chapter addresses a disturbance observer-based control (DOBC) approach for nonlinear systems under disturbances, that is, nonlinear DOBC or NDOBC. Within the NDOBC framework, instead of considering the control problem for a nonlinear system with disturbances as a single one, it is divided into two subproblems, each with its own design objectives. The first subproblem is the same as the control problem for a nonlinear system without disturbances and its objective is to stabilize the nonlinear plant and obtain desired tracking and/or regulation performance specifications. The second subproblem is disturbance attenuation. A nonlinear disturbance observer is designed to estimate external disturbances and then the estimated value is employed to compensate for the influence of disturbances. DOBC for linear systems has been developed and employed in engineering over three decades. Ohishi et al. [27] pioneered the development of DOBC for motion control systems. After that, DOBC has been employed in many mechatronic systems including disk drivers, machining centers, dc/ac motors, and manipulators [38, 81]. Most of the work on DOBC is engineering-oriented and lacks sound theoretic justification. When an attempt is made to extend DOBC from linear systems to nonlinear systems, this results in a composite controller consisting of a nonlinear controller and a nonlinear disturbance observer. Analysis and design of such a composite control system is challenging. Recent work has been concentrated on the development of nonlinear disturbance observers. Oh and Chung first improved a linear disturbance observer in robots using the information of nonlinear inertial coupling dynamics [23]. The application of this modified observer in redundant manipulators provides improved performance. A sliding mode-based nonlinear disturbance observer was proposed and applied in motor control in [85]. Chen et al. [2] developed a nonlinear disturbance observer for unknown constant disturbance using Lyapunov stability theory and applied it to a two-link manipulator.

This chapter aims to develop a more general framework for NDOBC and also establish rigorous basis for NDOBC development. The contents mainly refer to [5]. First, a general procedure for the design of DOBC for nonlinear systems is presented. This procedure is then applied to the control problem for a nonlinear system subject to disturbances generated by an exogenous system. This kind of disturbances widely exist in engineering systems including unknown load and harmonics, and has been investigated in several linear/nonlinear control approaches, most notably, nonlinear output regulation theory [107].

5.2 A General Design Framework

A nonlinear system is depicted by

$$\begin{cases} \dot{x} = f(x) + g_1(x)u + g_2(x)d, \\ y = h(x), \end{cases} \tag{5.1}$$

where $x \in R^n$, $u \in R$, $d \in R$, and $y \in R$ denote the state vector, the control input, the disturbance and the output, respectively. It is supposed that $f(x)$, $g_1(x)$, $g_2(x)$, and $h(x)$ are smooth functions in terms of x. A general NDOBC design procedure for system (5.1) is stated as follows

1. Design a nonlinear controller for system (5.1) to obtain stability and other performance specifications under the assumption that the disturbance is measurable.
2. Construct a nonlinear disturbance observer to estimate the disturbance.
3. Integrate the disturbance observer with the nonlinear controller by replacing the disturbance in the control law with its estimation by the disturbance observer.

The block diagram of the proposed NDOBC is shown by Figure 5.1. As shown by this figure, the composite controller consists of two parts: a controller without or having poor disturbance attenuation ability and a disturbance observer.

Remark 5.1 *The above design procedure consists of two stages. In the first stage, the controller is designed under the assumption that the disturbance is measurable. All the existing methods for designing a linear controller or a nonlinear controller can be used*

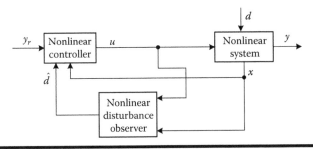

Figure 5.1 **Structure of nonlinear disturbance observer based control.**

in this stage. In the second stage, a linear/nonlinear disturbance observer is designed and then integrated with the previous designed controller.

5.3 Nonlinear Disturbance Observer-Based Control (NDOBC)

5.3.1 Nonlinear Disturbance Observer

Supposed that the disturbances are generated by the following exogenous system

$$\begin{cases} \dot{\xi} = A\xi, \\ d = C\xi, \end{cases} \tag{5.2}$$

where $\xi \in R^s$ and $d \in R$. In general, the exogenous system (5.2) is supposed to be neutrally stable, which indicates that a persistent disturbance is imposed on system (5.1).

In order to estimate the unknown disturbance d, a nonlinear disturbance observer is presented in [5], which is depicted by

$$\begin{cases} \dot{z} = [A - l(x)g_2(x)C]z + Ap(x) - l(x)[g_2(x)Cp(x) + f(x) + g_1(x)u], \\ \hat{\xi} = z + p(x), \\ \hat{d} = C\hat{\xi}. \end{cases} \tag{5.3}$$

Theorem 5.1 *Consider system (5.1) under the disturbances generated by exogenous system (5.2). Enhanced harmonic disturbance observer (5.3) can exponentially estimate the disturbances if nonlinear observer gain $l(x)$ is selected such that*

$$\dot{e} = [A - l(x)g_2(x)C]e, \tag{5.4}$$

is globally exponentially stable regardless of x.

Theorem 5.2 *The estimation $\hat{d}$ yielded by harmonic nonlinear disturbance observer (5.3) converges to the disturbance d globally exponentially if there exists a gain K such that transfer function*

$$H(s) = C(s I - \bar{A})^{-1} K, \tag{5.5}$$

is asymptotically stable and strictly positive real where

$$\bar{A} = (A - K\alpha_0 C). \tag{5.6}$$

The detailed proof procedures for Theorems 5.1 and 5.2 have been given in Chapter 3, which are omitted here for space.

5.3.2 Composite Controller Design

After the nonlinear disturbance observer is designed as (5.3), it is integrated with a separately designed controller. The stability of the closed-loop system under the composite controller as shown in Figure 5.1 is investigated in this section. The main result of this section is concluded by the following theorem.

Theorem 5.3 *Consider nonlinear system (5.1) under disturbance (5.2) with well-defined disturbance-to-output relative degree. The closed-loop system under the nonlinear composite controller (that is, NDOBC) designed by the procedure in Section 5.2, as shown in Figure 5.1, is semiglobally exponentially stable in the sense that for an initial state x and ξ satisfying*

$$||x(0)|| \leq R_1, ||\xi(0)|| \leq R_2,$$

where R_1 and R_2 are given scalars (could be arbitrarily large)

$$\lim_{t \to \infty} x(t) \to 0, \tag{5.7}$$

and

$$\lim_{t \to \infty} e(t) \to 0, \tag{5.8}$$

if the following conditions are satisfied: 1) when disturbance d is measurable, there exists a control law u(x, d) such that the closed-loop system is globally exponentially stable regardless of disturbance; 2) nonlinear disturbance observer (5.3) is designed with the chosen design function p(x)-and there exists a gain function K such that transfer function

$$H(s) = C(s I - \bar{A})^{-1} K \tag{5.9}$$

is asymptotically stable and strictly positive real.

Proof In order to asymptotically stabilize the system (5.2) regardless of disturbance, a part of the control effort, $u(x, d)$, shall linearly depend on disturbances d. To this end, the control law can be divided into

$$u(x, d) = \beta(x) + \gamma(x)d. \tag{5.10}$$

Substituting (5.10) into system (5.1) gives

$$\dot{x} = f(x) + g_1(x)\beta(x) + g_1(x)\gamma(x)d + g_2(x)d \tag{5.11}$$

which indicates that in order to exponentially stabilize system (5.1) under arbitrary disturbance, there must exists a $\gamma(x)$ such that

$$g_1(x)\gamma(x) = -g_2(x). \tag{5.12}$$

Under this condition, closed-loop system (5.11) reduces to

$$\dot{x} = f(x) + g_1(x)\beta(x), \tag{5.13}$$

which is globally exponentially stable under an appropriately designed $\beta(x)$.

Since disturbance d is unmeasurable, it is estimated by nonlinear disturbance observer (5.3). It is concluded from Theorem 5.2 that condition (2) in Theorem 5.3 indicates that nonlinear disturbance observer (5.3) with design function

$$p(x) = KL_f^{r-1}h(x)$$

is globally exponentially stable. After replacing disturbance d with its estimate in control law (5.10), according to condition (5.12), closed-loop system (5.11) becomes

$$\dot{x} = f(x) + g_1(x)\beta(x) + g_2(x)(d - \hat{d}) \tag{5.14}$$

Combining (5.14) with observer dynamics (5.4), the closed-loop system under the composite controller is obtained and described by

$$\begin{cases} \dot{x} = f(x) + g_1(x)\beta(x) + g_2(x)e_1, \\ \dot{e} = (A - l(x)g_2(x)C)e. \end{cases} \tag{5.15}$$

where $e_1 = d - \hat{d} = C(\xi - \hat{\xi}) = Ce$.

The semiglobal stability of the above composite system is shown as follows. Since system (5.13) is globally exponentially stable, there exists a Lyapunov function

$V_c(x)$ such that its derivative along system (5.13) satisfies

$$\dot{V}_c(x) = \frac{\partial V_c(x)}{\partial x}(f(x) + g_1(x)\beta(x)) < -\delta_1||x|| \tag{5.16}$$

where δ_1 is a small positive scalar. Taking

$$V(x, e) = V_c(x) + \mu V_o(e) = V_c(x) + \mu e^T P e \tag{5.17}$$

as a Lyapunov candidate for system (5.15) where μ is a large positive scalar to be determined, one obtains

$$\dot{V}_c(x) = \frac{\partial V_c(x)}{\partial x}(f(x) + g_1(x)\beta(x) + g_2(x)e_1) + 2\mu e^T(A - l(x)g_2(x)C)e. \tag{5.18}$$

The global exponential stability of disturbance observer error system indicates that

$$\dot{V}_o(e) < -\delta e^T e, \tag{5.19}$$

for any x and e.

Since transfer function (5.9) is asymptotically stable and strictly positive real, substitution of (5.19) and (5.16) into (5.18) gives

$$\begin{aligned}\dot{V}_c(x) &= \tfrac{\partial V_c(x)}{\partial x}(f(x) + g_1(x)\beta(x) + g_2(x)e_1) - \mu\delta e^T e \\ &< -\delta_1||x|| + \tfrac{\partial V_c(x)}{\partial x}g_2(x)e_1 - \mu\delta e^T e.\end{aligned} \tag{5.20}$$

In the same fashion as in [106], it is concluded that all the states and observer errors starting from a possible arbitrarily large set converge to the origin as $t \to \infty$.

5.4 Example Study

The effectiveness of the proposed NDOBC approach is demonstrated by controlling of a two-link robotic manipulator, each link of which is directly driven by a dc motor [2]. The dynamic model of manipulator including the first order dynamics of dc motors is given by

$$J(\theta(t))\ddot{\theta}(t) + G(\theta(t), \dot{\theta}(t)) = u(t) + d(t)' \tag{5.21}$$

where $\theta \in R^2, \dot{\theta} \in R^2$ are displacement and velocity of the links. $u(t) \in R^2$ is the voltage applied on the motors and d' the equivalent torque disturbances on the motor input. It is assumed that in this simulation study that there is only reaction torque from the object/environment acting on the tip of the second link in the operation. Hence, d' can be expressed as

$$d'(t) = c_2 d(t) \tag{5.22}$$

where

$$c_2 = \begin{bmatrix} 0 \\ 1 \end{bmatrix}.$$

According to the design procedure in Section 5.2, first it is assumed that there are no disturbances and the well-known computed torque control (CTC) method is employed to control the two link robotic manipulator. The resulted CTC law is given by

$$u^*(t) = -\left\{ J(\theta)(K_p(\theta - \theta_d) + K_v(\dot\theta - \dot\theta_d) - \ddot\theta_d) - G(\theta, \dot\theta) \right\}, \tag{5.23}$$

where θ_d denotes reference trajectory of the manipulator, and $\dot\theta_d$ and $\ddot\theta_d$ represent velocity and acceleration of the reference trajectory, respectively. The feedback gain matrices $K_p = \mathrm{diag}\{k_{p1}, k_{p2}\}$ and $K_v = \mathrm{diag}\{k_{v1}, k_{v2}\}$ are designed according to closed-loop performance requirements.

It is shown in Figures 5.2 and 5.3 by the dashed lines (which are almost indistinguishable from solid lines) that satisfactory performance can be achieved by CTC with feedback gains $k_{pi} = 30$, $k_{vi} = 7.5$, for $i = 1, 2$ in the absence of disturbance. It is suppose that there is a periodic disturbance torque acting on the end of the second link, given by

$$d(t) = 10\sin(2t + 1) N \cdot m.$$

As shown in Figures 5.2 and 5.3 by the dotted lines, the performance of the baseline CTC is significantly degraded, in particular for the second link.

A nonlinear disturbance observer is then designed to enhance the disturbance attenuation ability of the CTC by following the design procedure presented in Section 5.2. The disturbance $d(t)$ is generated by a neutrally stable system described by (5.2) with

$$A = \begin{bmatrix} 0 & 2 \\ -2 & 0 \end{bmatrix}, \quad C = [1 \; 0].$$

According to the above design procedure and guideline, nonlinear variable $p(x)$ is designed as

$$p(x) = K\dot\theta_2 \tag{5.24}$$

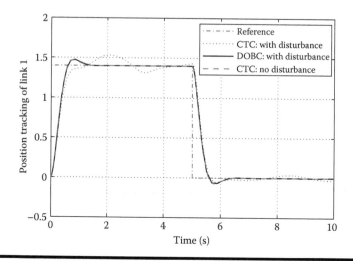

Figure 5.2 **Performance of disturbance observer enhanced computed torque controller: Link 1.**

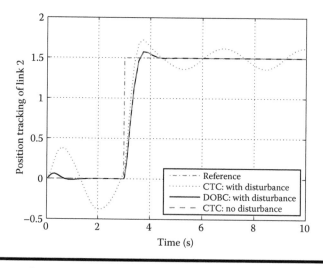

Figure 5.3 **Performance of disturbance observer enhanced computed torque controller: Link 2.**

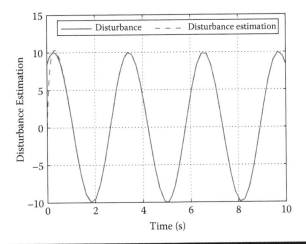

Figure 5.4 **Disturbance estimation by the proposed NDO.**

where

$$K = \begin{bmatrix} k_1 \\ k_2 \end{bmatrix},$$

is the gain matrix and designed as $k_1 = k_2 = 10$ in the simulation.

The above designed disturbance observer is then integrated with the CTC to enhance the disturbance attenuation ability. As shown in Figures 5.2 and 5.3 by the solid lines for Links 1 and 2, respectively, the performance of CTC under the disturbance is significantly improved and the NDOBC achieves good disturbance attenuation ability. There is a tracking error caused by disturbance in the second link at the beginning for the proposed NDOBC method. However, good tracking performance is achieved after a short period. The response curve of disturbance estimation for such system has been shown by Figure 5.4, which implies that the proposed NDO can effectively estimate the disturbance generated by exogenous system (5.2).

5.5 Summary

This chapter has proposed a general framework for design of controllers for non-linear systems with disturbances using DOBC techniques. It has been shown that this approach is quite flexible and can be integrated with current linear/nonlinear control methods that have poor disturbance attenuation ability. The effectiveness of the nonlinear DOBC procedure has been illustrated by the control problem of a class of nonlinear systems subject to disturbances generated by a linear exogenous system. Theoretic results have been obtained for the DOBC approach to this problem including the global exponential stability of nonlinear disturbance observer and the semiglobal exponential stability of composite controller.

Chapter 6

Generalized Extended State-Observer-Based Control for Systems with Mismatched Uncertainties

6.1 Introduction

During the past several decades, many elegant approaches have been proposed to estimate disturbances, including unknown input observer (UIO) [69], disturbance observer (DO) [27, 2, 5, 86], perturbation observer (POB) [70, 71], equivalent input disturbance (EID)-based estimation [72, 73], and extended state observer (ESO) [4]–[76]. The sliding-mode disturbance observer was also proposed for disturbance estimation using states and control inputs information in References [108, 109]. Note that all these methods are designed based on the model of plant. A natural doubt would be what least information does a designer have to know about the plant in order to build the estimator [74]? Among the above listed approaches, it is reported that ESO requires the least amount of plant information [110]; in fact, only the system order should be known. Due to such a promising feature, ESO-based control (ESOBC) schemes (also known as active disturbance rejection control (ADRC)) become more and more popular in recent years.

Although the ESOBC has obtained successful applications in many practical control systems, one factor that severely constrains the application of basic ESOBC

Table 6.1 Applicability of the proposed GESOBC and the ESOBC

Method	Disturbance type	Variable type	System type
GESOBC	Mismatched/matched	MIMO	Nonintegral-chain
ESOBC	Matched	SISO	Essential-integral-chain

method is that the uncertainties in many practical systems may not satisfy the so-called matching condition [100]. Here the matching condition implies that the uncertainties act via the same channel as that of the control input. In this chapter, a generalized extended state observer-based control (GESOBC) method is presented to solve the disturbance attenuation problem of a class of nonintegral-chain systems with mismatched uncertainties. It is shown that by properly designing a disturbance compensation gain, the mismatched uncertainties can be attenuated from system output. A systematic method is developed for the disturbance compensation gain design. Parameter design of the proposed method is discussed in detail, and feasible conditions for extending the proposed GESOBC to multi-input–multi-output (MIMO) systems without any coordinate transformations are also investigated. The GESOBC method largely extends the applicability of the ESOBC since it shows many superiorities over the basic ESOBC method, listed in Table 6.1, where the essential-integral-chain systems implies that the systems can be converted into integral chain systems by transformation.

An uncertain system with the order of n under the basic consideration is usually the following integral chain system

$$
\begin{cases}
\dot{x}_1 = x_2, \\
\dot{x}_2 = x_3, \\
\quad \vdots \\
\dot{x}_{n-1} = x_n, \\
\dot{x}_n = f(x_1, x_2, \ldots, x_n, d(t), t) + bu, \\
y = x_1.
\end{cases}
\tag{6.1}
$$

The basic ESO for system (6.1) has been generally designed as [9]

$$
\begin{cases}
\dot{z}_1 = z_2 - \beta_1(z_1 - y), \\
\dot{z}_2 = z_3 - \beta_2(z_1 - y), \\
\quad \vdots \\
\dot{z}_n = z_{n+1} - \beta_n(z_1 - y) + bu, \\
\dot{z}_{n+1} = -\beta_{n+1}(z_1 - y),
\end{cases}
\tag{6.2}
$$

The resultant basic ESOBC law is usually designed as [75, 4]

$$u = K_x x - \frac{z_{n+1}}{b},$$ (6.3)

where K_x is the feedback control gain. The basic ESOBC method is possibly unavailable for the following simple second-order system

$$\begin{cases} \dot{x}_1 = x_1 - 2x_2 + f(x_1, x_2, d(t), t), \\ \dot{x}_2 = x_1 + x_2 + u. \end{cases}$$ (6.4)

System (6.4) does not satisfy the basic formulation as (6.1) in the following two aspects. On the one hand, (6.4) does not satisfy the integral chain form. On the other hand, uncertainties $f(x_1, x_2, d(t), t)$ enter the system with a different channel from the control input u, that is, the so-called matching condition is not satisfied. For the above mentioned case, basic ESOBC law (6.3) is no longer available. Thus, it is imperative to develop generalized extended-state observer-based control (GESOBC) for general systems that do not satisfy the basic formulation of system (6.1). The contents of this chapter mainly refer to [14].

6.2 Generalized Extended-State Observer-Based Control (GESOBC)

For the sake of simplicity, the following single-input–single-output (SISO) system with mismatched uncertainties is taken into account

$$\begin{cases} \dot{x} = Ax + b_u u + b_d f(x, d(t), t), \\ y_m = C_m x, \\ y_o = c_o x, \end{cases}$$ (6.5)

where $x \in R^n$, $u \in R$, $d \in R$, $y_m \in R^r$, and $y_o \in R$ are the state vector, input, external disturbance, measurable outputs, and controlled output, respectively. $f(x, d(t), t)$ is the uncertain function in terms of x and d. A with dimension $n \times n$, b_u with dimension $n \times 1$, b_d with dimension $n \times 1$, C_m with dimension $r \times n$, and c_o with dimension $1 \times n$ are system matrices, respectively.

Remark 6.1 *In (6.5), function $f(x, d(t), t)$ denotes the lumped disturbance, which possibly include external disturbances, unmodeled dynamics, parameter variations, and complex nonlinear dynamics, which may be difficult for the feedback part to deal with.*

Remark 6.2 *Equation (6.5) denotes a more general class of systems as compared with system (2.26) since system (6.5) is not confined to integral chain form and may subject* to mismatched uncertainties and disturbances [111]. The matching case is a special case of (6.5), by simply taking $b_u = \lambda b_d$, $\lambda \in R$.

6.2.1 Composite Control Design

Similar to the basic case in Section 2.3, adding an extended variable

$$x_{n+1} = d = f(x, d(t), t), \tag{6.6}$$

to linearize system (6.5), the extended system equation is obtained

$$\begin{cases} \dot{\bar{x}} = \bar{A}\bar{x} + \bar{b}_u u + E h(t), \\ y_m = \bar{C}_m \bar{x}, \end{cases} \tag{6.7}$$

where variables

$$\bar{x} = \begin{bmatrix} x \\ x_{n+1} \end{bmatrix}, \quad h(t) = \frac{d f(x, d(t), t)}{dt},$$

and matrices

$$\bar{A} = \begin{bmatrix} A_{n\times n} & (b_d)_{n\times 1} \\ 0_{1\times n} & 0_{1\times 1} \end{bmatrix}_{(n+1)\times(n+1)}, \quad \bar{b}_u = \begin{bmatrix} (b_u)_{n\times 1} \\ 0_{1\times 1} \end{bmatrix}_{(n+1)\times 1},$$

$$E = \begin{bmatrix} 0_{n\times 1} \\ 1_{1\times 1} \end{bmatrix}_{(n+1)\times 1}, \quad \bar{C}_m = [C_m, \ 0_{r\times 1}]_{r\times(n+1)}.$$

Assumption 6.1 $(A, \ b_u)$ *is controllable, and* $(\bar{A}, \ \bar{C}_m)$ *is observable.*

Remark 6.3 *A necessary condition of* $(\bar{A}, \ \bar{C}_m)$ *observable is* $(A, \ C_m)$ *observable. The details are stated as follows.*

The observability matrices of (A, C_m) and $(\bar{A}, \bar{C}_m)$ are

$$P_o = \begin{bmatrix} C_m \\ C_m A \\ \vdots \\ C_m A^{n-1} \end{bmatrix}, \tag{6.8}$$

and

$$
\bar{P}_o =
\left[
\begin{array}{c|c}
C_m & 0 \\
C_m A & C_m b_d \\
C_m A^2 & C_m A b_d \\
\vdots & \vdots \\
C_m A^{n-1} & C_m A^{n-2} b_d \\
C_m A^n & C_m A^{n-1} b_d
\end{array}
\right].
\tag{6.9}
$$

Suppose that (A, C_m) is not observable. Without loss of generality, $C_m A^{n-1}$ in matrix P_o can be expressed as

$$
C_m A^{n-1} = \lambda_1 C_m + \lambda_2 C_m A + \cdots + \lambda_{n-1} C_m A^{n-2},
\tag{6.10}
$$

where $\lambda_i (i = 1, \ldots, n-1)$ are constant coefficients.

Combining (6.9) with (6.10), gives

$$
\begin{aligned}
\left[C_m A^n, \ C_m A^{n-1} b_d \right] &= \lambda_1 \left[C_m A, \ C_m b_d \right] + \lambda_2 \left[C_m A^2, \ C_m A b_d \right] \\
&\quad + \cdots + \lambda_{n-1} \left[C_m A^{n-1}, \ C_m A^{n-2} b_d \right].
\end{aligned}
\tag{6.11}
$$

It can be derived from (6.9) and (6.11) that $rank(\bar{P}_o) < n$. This implies the conclusion of Remark 6.3.

For system (6.7), a extended state observer is designed as follows

$$
\begin{cases}
\dot{\hat{x}} = \bar{A}\hat{x} + \bar{b}_u u + L(y_m - \hat{y}_m) \\
\hat{y}_m = \bar{C}_m \hat{x},
\end{cases}
\tag{6.12}
$$

where $\hat{x} = \left[\hat{x}^T, \ \hat{x}_{n+1} \right]^T$, $\hat{x}$, and $\hat{x}_{n+1}$ are the estimates of state variable $\bar{x}$, x, and x_{n+1} in (6.7), respectively. Matrix L with dimension $(n+1) \times r$ is the observer gain to be designed.

In the presence of mismatched uncertainties, the basic ESOBC law, $u = K_x x - \hat{d}$ (where $\hat{d} = \hat{x}_{n+1}$, K_x the feedback control gain), is unable to effectively compensate the uncertainties in system (6.5).

Remark 6.4 *It should be pointed out that the mismatched uncertainties can not be attenuated completely from the state equation no matter what controller is designed [106]. One of the most achievable goal in this case is to remove the uncertainties from the output channel in steady state.*

The composite control law is designed as

$$
u = K_x x + K_d \hat{d},
\tag{6.13}
$$

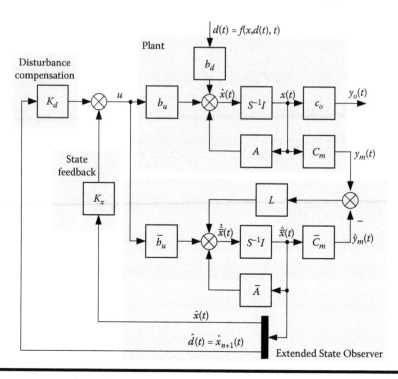

Figure 6.1   Configuration of the proposed GESOBC method.

or

$$u = K_x \hat{x} + K_d \hat{d}, \tag{6.14}$$

where K_x is the feedback control gain, and K_d is the disturbance compensation gain, designed as

$$K_d = -[c_o(A + b_u K_x)^{-1} b_u]^{-1} c_o(A + b_u K_x)^{-1} b_d. \tag{6.15}$$

Remark 6.5   *The disturbance compensation gain K_d in (6.15) is a general case and suitable for both matching and mismatching cases. For the matching case, i.e., $b_u = \lambda b_d$, $\lambda \in R$, it can be obtained from (6.15) that the disturbance compensation gain reduces to $K_d = -1/\lambda$, which is the same as the basic ESOBC law (6.3) in previous literatures.*

The configuration of the proposed generalized extended state observer based control is shown by Figure 6.1.

6.2.2 Stability and Disturbance Rejection Analysis

Assumption 6.2 *The lumped disturbances satisfy the following conditions: (1) $d(t) = f(x, d(t), t) = \bar{f}(d(t), t)$; (2) $d(t)$ is bounded; (3) $d(t)$ has a constant value in steady state, that is, $\lim\limits_{t\to\infty} \dot{d}(t) = \lim\limits_{t\to\infty} h(t) = 0$ and $\lim\limits_{t\to\infty} d(t) = D_c$.*

The estimation errors of state and disturbance are defined as

$$e_x = \hat{x} - x, \tag{6.16}$$

$$e_d = \hat{d} - d, \tag{6.17}$$

where $\hat{d} = \hat{x}_{n+1}$ represents the estimate of system uncertainties. Combining Equations (6.7), (6.12), (6.16) with (6.17), the estimation error equation is governed by

$$\dot{e} = A_e e - E h(t), \tag{6.18}$$

where

$$e = \begin{bmatrix} e_x \\ e_d \end{bmatrix}, \quad A_e = \bar{A} - L\bar{C}_m. \tag{6.19}$$

The boundedness stability of ESO has been concluded by the following lemma.

Lemma 6.1 *[112] Assuming that the observer gain vector L in (6.12) is chosen such that A_e is a Hurwitz matrix, then the observer error, e for the ESO is bounded for any bounded $h(t)$.*

Lemma 6.2 *[14] The following single-input linear system*

$$\dot{x} = Ax + Bu, \tag{6.20}$$

is asymptotically stable if A is a Hurwitz matrix, u is bounded and satisfies $\lim\limits_{t\to\infty} u(t) = 0$.

Lemma 6.3 *For system (6.20), if matrix A is Hurwitz and $\lim\limits_{t\to\infty} u(t) = U_c \neq 0$, the states converge to a constant vector $-A^{-1}BU_c$, that is, $\lim\limits_{t\to\infty} x(t) = -A^{-1}BU_c$. The result can easily follow from Lemma 6.2 by coordinate transformations.*

6.2.2.1 Case of Measurable States

If the states are available, the composite control law is designed as (6.13). The stability and disturbance rejection performance is analyzed by the following theorems.

Theorem 6.1 *Suppose that Assumption 6.1 is satisfied. The bounded stability of system (6.5) under the proposed GESOBC law (6.13) for any bounded $h(t)$ and $d(t)$ is guaranteed if the observer gain L in (6.12) and the feedback control gain K_x in (6.13) are selected such that A_e in (6.19) and $A_f = A + b_u K_x$ are Hurwitz matrices, respectively.*

Proof Combining system (6.5), composite control law (6.13), with error system (6.18), the closed-loop system is governed by

$$
\begin{bmatrix} \dot{x} \\ \dot{e} \end{bmatrix} = \begin{bmatrix} A_f & b_u \bar{K} \\ 0 & A_e \end{bmatrix} \begin{bmatrix} x \\ e \end{bmatrix} + \begin{bmatrix} 0 & b_d + b_u K_d \\ -E & 0 \end{bmatrix} \begin{bmatrix} h \\ d \end{bmatrix},
\tag{6.21}
$$

where $\bar{K} = [0_{1 \times n}, \ K_d]$. Considering the condition that both A_f and A_e are Hurwitz matrices yields

$$
\begin{bmatrix} A_f & b_u \bar{K} \\ 0 & A_e \end{bmatrix},
$$

which is also a Hurwitz matrix.

It can be concluded from Lemma 6.1 that closed-loop system (6.27) is bounded-input-bounded-output (BIBO) stable for any bounded $h(t)$ and $d(t)$ if K and L are appropriately chosen.

Theorem 6.2 *Suppose that Assumptions 6.1 and 6.2 are satisfied, and also the observer gain L and the feedback control gain K_x are chosen such that matrices A_e in (6.19), A_f Hurwitz, and $c_o A_f^{-1} b_u$ invertible. For system (6.5) under control law (6.13), the lumped disturbances can be attenuated from the output channel in steady state under the proposed GESOBC law (6.13).*

Proof Substituting control law (6.13) into system (6.5) and using (6.17), the state is expressed as

$$
x = (A + b_u K_x)^{-1} [\dot{x} - b_u K_d e_d - (b_u K_d + b_d) d].
\tag{6.22}
$$

Combining (6.5), (6.15), with (6.22), gives

$$
y_o = c_o (A + b_u K_x)^{-1} \dot{x} + c_o (A + b_u K_x)^{-1} b_d e_d.
\tag{6.23}
$$

It can be observed from (6.23) that the lumped disturbances are removed from the output channels. Under the given conditions, it is obtained from Lemmas 6.2 and 6.3, and Theorem 6.1 that

$$
\lim_{t \to \infty} \dot{x}(t) = 0, \ \lim_{t \to \infty} e(t) = \mathbf{0}.
\tag{6.24}
$$

Combining (6.23) with (6.24) gives

$$\lim_{t \to \infty} y_o(t) = 0. \tag{6.25}$$

6.2.2.2 Case of Unmeasurable States

If the state variables are unmeasurable, the estimate of both the lumped disturbances and the states are available for control design. The composite control law is designed as (6.14) in this case. Let $K = [K_x, \ K_d]$, a compact expression of control law (6.14) is obtained

$$u = K\hat{x}. \tag{6.26}$$

Theorem 6.3 *Suppose that Assumption 6.1 is satisfied. The bounded stability of system (6.5) under the proposed GESOBC law (6.14) for any bounded $h(t)$ and $d(t)$ is guaranteed if the observer gain L in (6.12) and the feedback control gain K_x in (6.14) are selected such that A_e and A_f are Hurwitz matrices, respectively.*

Proof Combining system (6.5), composite control law (6.14) with error system (6.18), the closed-loop system is expressed as

$$\begin{bmatrix} \dot{x} \\ \dot{e} \end{bmatrix} = \begin{bmatrix} A_f & b_u K \\ 0 & A_e \end{bmatrix} \begin{bmatrix} x \\ e \end{bmatrix} + \begin{bmatrix} 0 & b_d + b_u K_d \\ -E & 0 \end{bmatrix} \begin{bmatrix} h \\ d \end{bmatrix}. \tag{6.27}$$

Since both A_f and A_e are Hurwitz matrices, it is easy to show that matrix

$$\begin{bmatrix} A_f & b_u K \\ 0 & A_e \end{bmatrix},$$

is also a Hurwitz matrix. The proof is completed by using the result in Lemma 6.1.

Theorem 6.4 *Suppose that Assumptions 6.1 and 6.2 are satisfied, and also the observer gain L and the feedback control gain K_x are designed such that matrices A_e in (6.19), A_f Hurwitz, and $c_o A_f^{-1} b_u$ are invertible. For system (6.5) under control law (6.14), the lumped disturbances can be removed from the output channel in steady state with the proposed GESOBC law (6.14).*

Proof Combining (6.5) with (6.14) yields

$$x = (A + b_u K_x)^{-1} [\dot{x} - b_u K e - (b_u K_d + b_d) d]. \tag{6.28}$$

Based on (6.5), (6.28) and (6.15), the output is expressed as

$$y_o = c_o (A + b_u K_x)^{-1} (\dot{x} - b_u K_x e_x) + c_o (A + b_u K_x)^{-1} b_d e_d. \tag{6.29}$$

Considering (6.24), the same result as (6.25) can be obtained from (6.29).

6.3 Simulation Example

To demonstrate the efficiency of the proposed GESOBC approach, a second-order uncertain nonlinear system subject to mismatched disturbances is considered

$$
\begin{cases}
\dot{x}_1 = x_2 + e^{x_1} + w, \\
\dot{x}_2 = -2x_1 - x_2 + u, \\
y = x_1.
\end{cases}
\tag{6.30}
$$

Let

$$
A = \begin{bmatrix} 0 & 1 \\ -2 & -1 \end{bmatrix}, \; b_u = \begin{bmatrix} 0 \\ 1 \end{bmatrix}, \; b_d = \begin{bmatrix} 1 \\ 0 \end{bmatrix},
$$

$$
C_m = c_o = [1 \; 0], \; f(x, w(t), t) = e^{x_1} + w,
$$

it can be observed that system (6.30) has the formulation of (6.5).

To guarantee the convergence of ESO, the observer gain vector in (6.12) is designed as $L = \begin{bmatrix} 14 & -66 & 125 \end{bmatrix}^T$ such that the corresponding ESO poles are $p_{eso} = \begin{bmatrix} -5 & -5 & -5 \end{bmatrix}^T$. The feedback control gain in this example is designed as $K_x = \begin{bmatrix} -4 & -4 \end{bmatrix}$. The poles of the closed-loop system, regardless of disturbances and uncertainties, are $p_{cl} = \begin{bmatrix} -2 & -3 \end{bmatrix}^T$ under such feedback control gain. The disturbance compensation gain can be calculated by (6.15), giving as $K_d = -5$. Considering that the states are unmeasurable, composite control law (6.14) is employed. The initial states of system (6.30) are $x_0 = \begin{bmatrix} 0 & 2 \end{bmatrix}^T$. External disturbance $d = 3$ acts on the system at $t = 6$ sec. The control objective is to remove the uncertainties from the output channel. The tracking command of the output is zero during the simulation. The response curves of the real estimated states and their estimate errors are shown in the Figures 6.2–6.4. The corresponding time history of the control input is shown by Figure 6.5.

It can be observed from Figure 6.2 that the output converges to its setpoint value quickly in the presence of both uncertainties and external disturbances. As shown in Figures 6.2–6.4, estimation errors of ESO converge to zero for all the states in such case of uncertainties.

6.4 Further Discussions

6.4.1 Extension to MIMO System

For the purpose of comparison with the basic ESOBC, only a SISO system with uncertainties in single channel is discussed in Section 6.2. Actually, the proposed GESOBC method can be extended to MIMO system with almost no modification. Here, the MIMO system may include multiple disturbances in different channels.

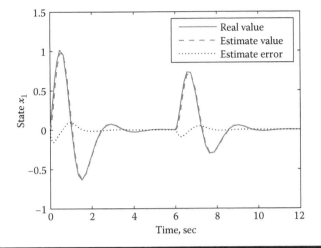

Figure 6.2   Response curves of real and estimated values of state x_1.

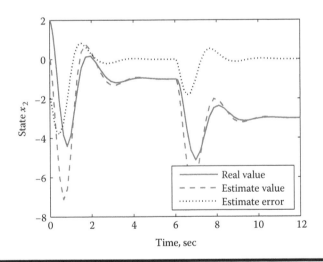

Figure 6.3   Response curves of real and estimated values of state x_2.

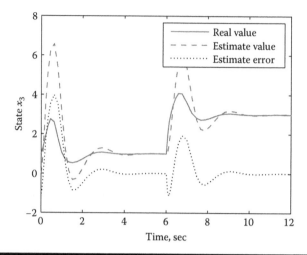

Figure 6.4 **Response curves of real and estimated disturbance $x_3 = d$.**

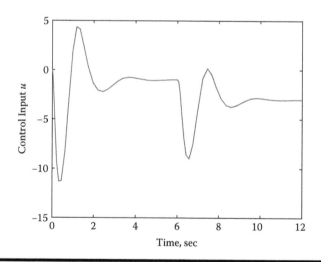

Figure 6.5 **Time history of the control input u.**

A general MIMO system is described as

$$\begin{cases} \dot{x} = Ax + B_u u + B_d f(x, d(t), t), \\ y_m = C_m x, \\ y_o = C_o x, \end{cases} \tag{6.31}$$

where $x \in R^n$, $u \in R^m$, $y_m \in R^r$, $y_o \in R^p$, and $f \in R^q$.

6.4.1.1 Solvability of the Disturbance Compensation Gain

The disturbance compensation gain in (6.15) is no longer available since $C_o(A + B_u K_x)^{-1} B_u$ is possibly noninvertible or even not a square matrix. In this case, it can be verified that an alternative but more general condition

$$C_o(A + B_u K_x)^{-1} B_u K_d = -C_o(A + B_u K_x)^{-1} B_d, \tag{6.32}$$

must be satisfied so as to guarantee feasibility of the proposed method.

Disturbance compensation gain K_d is able to be solved from (6.32) if the following rank condition holds

$$\text{rank}\,(C_o(A + B_u K_x)^{-1} B_u)) = \text{rank}\,([C_o(A + B_u K_x)^{-1} B_u,$$
$$- C_o(A + B_u K_x)^{-1} B_d]. \tag{6.33}$$

6.4.1.2 Controllable Condition

Another factor that possibly influences the feasibility of the proposed GESOBC for MIMO systems is the condition in Assumption 6.1 may not be satisfied with the increased number of the lumped disturbances. Consider a system with multiple lumped disturbances, depicted by

$$\begin{cases} \dot{x}_1 = x_2 + f_1(x_1, x_2, d, t), \\ \dot{x}_2 = -2x_1 - x_2 + u + f_2(x_1, x_2, d, t), \\ y_m = x_1. \end{cases} \tag{6.34}$$

It can be easily verified that $(\bar{A}, \bar{C}_m)$ is not observable for (6.34), and the proposed GESOBC is unavailable. However, the problem becomes feasible as more output information is accessible, which is shown by the following example that similar with (6.34) but with more measurable outputs

$$\begin{cases} \dot{x}_1 = x_2 + f_1(x_1, x_2, d, t), \\ \dot{x}_2 = -2x_1 - x_2 + u + f_2(x_1, x_2, d, t), \\ y_{1m} = x_1, \\ y_{2m} = x_2. \end{cases} \tag{6.35}$$

It can be verified that $(\bar{A}, \bar{C}_m)$ is observable for (6.35) now.

Generally speaking, besides the conditions in Assumption 6.1, condition (6.33) should be satisfied to guarantee feasibility of the GESOBC for MIMO system. If $(\bar{A}, \bar{C}_m)$ is not observable, one may make the problem feasible by seeking more output information.

6.4.2 Parameter Design for GESOBC

As for the proposed GESOBC method, there are mainly three parameters, including feedback control gain K_x, observer gain L, and disturbance compensation gain K_d, to be designed. A fixed way to determine K_d has been given in (6.15).

The most important designing parameters are K_x and L. As discussed in Theorems 6.1-6.4, the necessary conditions are L and K_x have to be designed such that the stability of the closed-loop system is guaranteed. However, these conditions are not sufficient. The reason is that the lumped disturbances would be a function of the states, which can only be estimated if the observer dynamics are faster than the closed-loop dynamics. The same argument for the state observer based control method is available.

It can be found from Section 6.2 that the poles of ESO and the closed-loop system are eigenvalues of matrix $A_e = \bar{A} - L\bar{C}_m$ and $A_f = A + b_u K_x$, respectively. If (A, b_u) is controllable and $(\bar{A}, \bar{C}_m)$ is observable, the poles of both the closed-loop system and ESO can be located arbitrarily.

To make the observer dynamic quicker than that of the closed-loop system, poles of ESO should be placed much farer away from the origin than those of the closed-loop system.

6.5 Summary

The basic extended state observer based control (ESOBC) method is only available for a class of SISO essential-integral-chain systems subject to disturbances/uncertainties satisfying the so-called matching condition. By appropriately designing a disturbance compensation gain, a GESOBC approach is employed for the control design of general systems with mismatched uncertainties and nonintegral-chain form. The GESOBC method can be extended to multi-input–multi-output (MIMO) systems with almost no modification.

Chapter 7

Nonlinear Disturbance Observer-Based Control for Systems with Mismatched Uncertainties

7.1 Introduction

A nonlinear disturbance observer-based robust control (NDOBRC) method is presented in this chapter to solve the disturbance attenuation problem of nonlinear systems with mismatched disturbances and uncertainties. In the presence of mismatched disturbances/uncertainties, it is usually impossible to achieve asymptotic stability as the disturbance and the control inputs do not appear in the same channels, and the influence of the disturbances can not be completely canceled. The design objective in the presence of mismatched disturbances is to remove the influence of disturbances and uncertainties from the output. The key issue here is how to design a disturbance compensation gain so as to eliminate the mismatched disturbances from the output channels completely. With the NDOBRC method the prominent "patch" feature retains, the robustness and disturbance attenuation against a much wider range of uncertainties and disturbances are significantly improved without sacrificing the nominal performance of the employed feedback control strategy. This chapter mainly refers to [15].

7.2 Problem Formulation

Consider a general single-input–single-output (SISO) affine nonlinear systems with lumped disturbances, represented as

$$\begin{cases} \dot{x} = f(x) + g_1(x)u + g_2(x)d, \\ y = h(x), \end{cases} \tag{7.1}$$

where $x \in R^n$, $d \in R^n$, $u \in R$ and $y \in R$ are the state vector, the lumped disturbance vector, and the input and the output variables. $f(x)$, $g_1(x)$, $g_2(x)$, and $h(x)$ are assumed to be smooth functions in terms of states x. Suppose that the lumped disturbances in system (7.1) satisfy the following assumption.

Assumption 7.1 *The lumped disturbances d are slowly time-varying, that is, $\dot{d}(t) \approx 0$.*

For system (7.1), the following nonlinear disturbance observer (NDO) is proposed in [40, 5, 2] to estimate the unknown disturbances d, given by

$$\begin{cases} \hat{d} = z + p(x), \\ \dot{z} = -l(x)g_2(x)z - l(x)[g_2(x)p(x) + f(x) + g_1(x)u], \end{cases} \tag{7.2}$$

where $\hat{d}$ and z are the estimates of the unknown disturbances and the internal states of nonlinear observer, respectively, and $p(x)$ is a nonlinear vector-valued function to be designed. Nonlinear observer gain $l(x)$ is defined as

$$l(x) = \frac{\partial p(x)}{\partial x}. \tag{7.3}$$

It can be proved that, under the assumption that the disturbances are slowly time-varying, $\hat{d}$ approaches d asymptotically if $p(x)$ is chosen such that

$$\dot{e}(t) + \frac{\partial p(x)}{\partial x}g_2(x)e(t) = 0, \tag{7.4}$$

is globally stable for all $x \in R^n$, where the estimation error is defined as $e = d - \hat{d}$. Any nonlinear vector-valued function $l(x)$ which makes (7.4) asymptotically stable can be chosen to guarantee the asymptotic convergence of estimation error.

Remark 7.1 *The rigorous asymptotic convergence of NDO has been established under the condition that the disturbances vary slowly relative to the observer dynamics (i.e., Assumption 7.1). However, it has been claimed in [2] that observer (7.2) can track some fast time-varying disturbances with bounded error as long as the derivative of disturbances is bounded.*

Remark 7.2 *In the presence of uncertainties, the lumped disturbances would be a function of the states, which can be reasonably estimated if the dynamics of disturbance observer are faster than that of the closed-loop dynamics. This can be referred to the same argument for the state observer based control methods.*

In the previous literature, the DOBC methods only handle the case of matched disturbances for nonlinear systems, that is, lumped disturbances d enter the system with the same channels as the control inputs. Such matching condition has largely restricted the application of DOBC approaches to more general controlled plants.

Remark 7.3 *NDO (7.2) is applicable to the case of mismatched disturbances. However, the estimates of NDO can not be used to compensate the disturbances directly since the disturbances are not in the same channels with the control inputs.*

In general, the influence of mismatched disturbances can not be removed from state variables. Based on the disturbance estimate of NDO (7.2), a composite control law as $u = \alpha(x) + \beta(x)\hat{d}$ is designed in this chapter to remove the influence of the lumped disturbance from the output channel by appropriately designing a compensation gain $\beta(x)$. This will largely extend the application scope of the DOBC approaches.

A general design procedure of the newly proposed DOBC for system (7.1) suffering from mismatched disturbances is proposed as follows:

(i) Design a baseline nonlinear feedback controller to achieve stability and performance specifications without considering disturbances/uncertainties.
(ii) Lump the external disturbances and the influences of the uncertainties, and then design a nonlinear disturbance observer to estimate the lumped disturbances.
(iii) Design a disturbance compensation gain so as to achieve desired performance in the presence of external disturbances and uncertainties.
(iv) Construct a composite NDOBRC law by integrating the nonlinear feedback controller and the disturbance observer-based compensation part.

7.3 Novel Nonlinear Disturbance Observer-Based Control

7.3.1 Controller Design

For nonlinear systems (7.1) with mismatched disturbances, a composite control law of NDOBRC is designed as

$$u = \alpha(x) + \beta(x)\hat{d}, \tag{7.5}$$

where $\alpha(x)$ is the feedback control law, $\beta(x)$ is the disturbance compensation gain vector to be designed, and $\hat{d}$ is the disturbance estimate by NDO (7.2).

In the composite control law (7.5), disturbance compensation term $\beta(x)\hat{d}$ is just designed for disturbances, that is, the NDO works if and only if disturbances exist. Thus it just works like a "patch" for the existing controller to enhance its disturbance attenuation and robustness against uncertainties. In the absence of disturbances and uncertainties, the closed-loop system shows its nominal control performance.

7.3.2 Stability Analysis

To prove the stability of closed-loop system, some definition and preliminary result are required, which are stated as follows.

Definition 7.1 [87] *A nonlinear system $\dot{x} = f(t, x, w)$ with $f : [0, \infty) \times R^n \times R^m \rightarrow R^n$ piecewise continuous in t and locally Lipschitz in x and w is said to be locally input-to-state stable (ISS) if there exist a class $\mathcal{KL}$ function χ, a class $\mathcal{K}$ function κ, and positive constants k_1 and k_2 such that for any initial state $x(t_0)$ with $\|x(t_0)\| < k_1$ and any input $w(t)$ with $\sup_{t \geq t_0} \|w(t)\| < k_2$, the solution $x(t)$ exists and satisfies*

$$\|x(t)\| \leq \chi(\|x(t_0)\|, t - t_0) + \kappa(\sup_{t_0 \leq \tau \leq t} \|w(t)\|)$$

for all $t \geq t_0 \geq 0$.

Lemma 7.1 *Consider system $\dot{x} = F(x, w) = f(x) + p(x)w$ with $F : D \times D_w \rightarrow R^n$ is locally Lipschitz in x, $D \subset R^n$ is a domain that contains $x = 0$, and $D_w \subset R^m$ is a domain that contains $w = 0$. If $p(x)$ is continuously differentiable and the origin of system $\dot{x} = f(x)$ is asymptotically stable, then $\dot{x} = F(x, w)$ is locally ISS.*

Proof According to Converse Lyapunov Theorem [87], it follows from the asymptotical stability of system $\dot{x} = f(x)$ that there exists a Lyapunov function $V(x)$ such that

$$\alpha_1(\|x\|) \leq V(x) \leq \alpha_2(\|x\|), \tag{7.6}$$

$$\frac{\partial V}{\partial x} f(x) \leq -\alpha_3(\|x\|), \tag{7.7}$$

$$\|\frac{\partial V}{\partial x}\| \leq \alpha_4(\|x\|), \tag{7.8}$$

in some bounded region around $x = 0$, where $\alpha_i(\cdot)(i = 1, 2, 3, 4)$ are class $\mathcal{K}$ functions. It follows from the continuous differentiable condition of $p(x)$ that $p(x)$

is bounded in the neighborhood Ω round $(x, w) = (0, 0)$, which implies that there exists a constant $L' \geq 0$ such that

$$\| p(x) \| \leq L', \forall (x, w) \in \Omega. \tag{7.9}$$

Taking derivative of the Lyapunov function $V(x)$ along the trajectory of system $\dot{x} = F(x, w)$, gives

$$
\begin{aligned}
\dot{V}(x) &= \frac{\partial V}{\partial x} F(x, w) \\
&= \frac{\partial V}{\partial x} f(x) + \frac{\partial V}{\partial x} p(x) w \tag{7.10} \\
&\leq -\alpha_3(\|x\|) + \alpha_4(\|x\|) L' \|w\|.
\end{aligned}
$$

Let $c > 0$ such that

$$\alpha_4(\|x\|) \leq c, \ \forall (x, w) \in \Omega. \tag{7.11}$$

Combining (7.10) with (7.11), gives

$$
\begin{aligned}
\dot{V}(x) &\leq -\alpha_3(\|x\|) + cL' \|w\| \\
&= -(1-\theta)\alpha_3(\|x\|) - \theta\alpha_3(\|x\|) + cL'\|w\| \tag{7.12} \\
&\leq -(1-\theta)\alpha_3(\|x\|), \ \forall \|x\| \geq \alpha_3^{-1}\left(\frac{cL'}{\theta}\|w\|\right),
\end{aligned}
$$

for $(x, w) \in \Omega$, where $0 < \theta < 1$. It followed from Theorem 5.2 in [113] that system $\dot{x} = f(x) + p(x)w$ is locally ISS.

Consider lumped disturbances d as the inputs of the closed-loop system, also the system states x and observer states e as the state of the closed-loop system. Let

$$\bar{x} = \begin{bmatrix} x \\ e \end{bmatrix}, \tag{7.13}$$

$$F(\bar{x}) = \begin{bmatrix} f(x) + g_1(x)\alpha(x) - g_1(x)\beta(x)e \\ -l(x)g_2(x)e \end{bmatrix}. \tag{7.14}$$

The input-to-state stability of closed-loop system is established by the following theorem.

Theorem 7.1 *The closed-loop system consists of nonlinear system (7.1), composite control law (7.5) and nonlinear disturbance observer (7.2) is ISS if the following conditions are satisfied:*

(i) Nonlinear system (7.1) under the original designed controller $u = \alpha(x)$ is globally asymptotically stable in the absence of disturbances.

(ii) *The vector-valued function $p(x)$ is chosen such that observer error system (7.4) is globally asymptotically stable.*

(iii) *There exists a disturbance compensation gain $\beta(x)$ such that $G(x) = g_2(x) + g_1(x)\beta(x)$ is continuously differentiable.*

Proof Combining system (7.1), composite control law (7.5), and disturbance estimation error function (7.4) together, the closed-loop system is obtained

$$\begin{cases} \dot{x} = [f(x) + g_1(x)\alpha(x)] - g_1(x)\beta(x)e + [g_2(x) + g_1(x)\beta(x)]d, \\ \dot{e} = -\dfrac{\partial p(x)}{\partial x} g_2(x)e. \end{cases} \tag{7.15}$$

Combining Equations (7.13), (7.14) with (7.15), the closed-loop system is given as

$$\dot{\bar{x}} = F(\bar{x}) + \begin{bmatrix} G(x) \\ 0 \end{bmatrix} d. \tag{7.16}$$

With the conditions given in (i) and (ii), it can be shown that $\dot{\bar{x}} = F(\bar{x})$ is asymptotically stable. Consider the condition given in (iii), it follows from Lemma 6.1 that the closed-loop system (7.16) is locally ISS.

7.3.3 Disturbance Attenuation Analysis

Suppose that there exist some nonsingular nonlinear functions $\bar{f}(x)$, $\bar{h}(x)$ and $\bar{\alpha}(x)$ such that $f(x) = \bar{f}(x)x$, $h(x) = \bar{h}(x)x$ and $\alpha(x) = \bar{\alpha}(x)x$. Nonlinear systems (7.1) and composite control law (7.5) can be rewritten as

$$\begin{cases} \dot{x} = \bar{f}(x)x + g_1(x)u + g_2(x)d, \\ y = \bar{h}(x)x, \end{cases} \tag{7.17}$$

and

$$u = \bar{\alpha}(x)x + \beta(x)\hat{d}. \tag{7.18}$$

Theorem 7.2 *Suppose Assumption 6.1 is satisfied. Consider nonlinear system (7.1) under composite control law (7.5) consisting of nonlinear feedback control law $\alpha(x)$ and disturbance compensation term $\beta(x)\hat{d}$ based on the estimates of NDO (7.2). The influence of lumped disturbances can be eliminated from the output channel in steady-state if the nonlinear disturbance compensation gain $\beta(x)$ is selected such that (i) closed-loop system (7.15) is ISS; (ii) following condition holds*

$$\beta(x) = -\left\{ \bar{h}(x) \left[\bar{f}(x) + g_1(x)\bar{\alpha}(x) \right]^{-1} g_1(x) \right\}^{-1} \times \bar{h}(x) \left[\bar{f}(x) \right.$$

$$\left. + g_1(x)\bar{\alpha}(x) \right]^{-1} g_2(x). \tag{7.19}$$

Proof Taking into account the closed-loop system (7.15) with Equations (7.17) and (7.18), the states can be expressed as

$$x = \left[\bar{f}(x) + g_1(x)\bar{\alpha}(x) \right]^{-1} \{ \dot{x} - g_1(x)\beta(x)e - [g_1(x)\beta(x) + g_2(x)] d \}. \quad (7.20)$$

Combining Equations (7.19), (7.20) with the output equation in (7.17) gives

$$y = \bar{h}(x) \left[\bar{f}(x) + g_1(x)\bar{\alpha}(x) \right]^{-1} \dot{x} + \bar{h}(x) \left[\bar{f}(x) + g_1(x)\bar{\alpha}(x) \right]^{-1} g_2(x)e. \tag{7.21}$$

Since the closed-loop system is stable, the following two conditions are satisfied, i.e., $\lim_{t \to \infty} \dot{x}(t) \to 0$ and $\lim_{t \to \infty} e(t) \to 0$. The later condition follows from the properly designed disturbance observers. It can be shown that the disturbances can be finally attenuated from the output in steady-state, that is, $y = 0$.

Remark 7.4 *Let $g_2(x) = (g_{21}(x), g_{22}(x), \ldots, g_{2n}(x))$, a more explicit expression of the disturbance compensation gain can be given as*

$$\beta(x) = (\beta_1(x), \beta_2(x), \ldots, \beta_n(x)), \tag{7.22}$$

where

$$\beta_i(x) = -\frac{\det \begin{bmatrix} \bar{f}(x) + g_1(x)\bar{\alpha}(x) & g_{2i}(x) \\ -\bar{h}(x) & 0 \end{bmatrix}}{\det \begin{bmatrix} \bar{f}(x) + g_1(x)\bar{\alpha}(x) & g_1(x) \\ -\bar{h}(x) & 0 \end{bmatrix}}.$$

Remark 7.5 *Nonlinear disturbance compensation gain $\beta(x)$ in (7.19) is a general case and suitable for both matched and mismatched disturbances. Actually, in matched case, that is, $g_1(x) = g_2(x)$, it can be obtained from (7.19) that the nonlinear disturbance compensation gain reduces to $\beta(x) = -1$ which is widely used in the previous DOBC designs [40, 86, 97, 93].*

7.4 Application to A Nonlinear Missile

7.4.1 Longitudinal Dynamics of A Missile System

The model of the longitudinal dynamics of a missile under consideration is taken from [40, 114, 115, 116], described by

$$\dot{\alpha} = f_1(\alpha) + q + b_1(\alpha)\delta + d_1, \tag{7.23}$$
$$\dot{q} = f_2(\alpha) + b_2\delta + d_2, \tag{7.24}$$

where α is the angle of attack (degrees), q is the pitch rate (degrees per second), and δ is the tail fin deflection (degrees). Disturbances d_1 and d_2 denote the lumped disturbance torques which may be caused by unmodeled dynamics, external wind, and variation of aerodynamic coefficients, etc. Nonlinear functions $f_1(\alpha)$, $f_2(\alpha)$, $b_1(\alpha)$, and b_2 are determined by aerodynamic coefficients. When the missile travels at Mach 3 at an altitude of 6,095 m (20,000 ft) and the angle of attack $|\alpha| \leq 20$ deg, they are given by [40, 115]

$$f_1(\alpha) = \frac{180g \, QS}{\pi \, WV} \cos(\frac{\pi \alpha}{180})(1.03 \times 10^{-4}\alpha^3 - 9.45 \times 10^{-3}\alpha|\alpha|$$

$$- 1.7 \times 10^{-1}\alpha), \tag{7.25}$$

$$f_2(\alpha) = \frac{180 \, QSd}{\pi \, I_{yy}}(2.15 \times 10^{-4}\alpha^3 - 1.95 \times 10^{-2}\alpha|\alpha| + 5.1 \times 10^{-2}\alpha), \tag{7.26}$$

$$b_1(\alpha) = -3.4 \times 10^{-2}\frac{180g \, QS}{\pi \, WV}\cos(\frac{\pi \alpha}{180}), \tag{7.27}$$

$$b_2 = -0.206\frac{180 \, QSd}{\pi \, I_{yy}}, \tag{7.28}$$

where parameters Q, S, W, V, d, and I_{yy} denote dynamic pressure, reference area, mass, velocity, reference diameter, and pitch moment of inertial of the missile system, respectively.

The tail fin actuator dynamics are approximated by a first-order lag process, that is,

$$\dot{\delta} = (1/t_1)(-\delta + u) + d_3, \tag{7.29}$$

where u the commanded fin defection (degrees), d_3 the disturbance which may influence the actuator dynamics (e.g., frictions) and t_1 time constant (seconds). The physical meaning and values of the parameters in Equations (7.25)–(7.29) for the missile under consideration are listed as follows: mass $W = 4,410$ kg, velocity $V = 947.6$ m/s, pitch moment of inertia $I_{yy} = 247.44$ kg·m^2, dynamic pressure $Q = 293,638$ N/m^2, reference area $S = 0.04087$ m^2, reference diameter $d = 0.229$ m, gravitational acceleration $g = 9.8$ m/s^2, and time constant of tin actuator $t_1 = 0.1$ s.

7.4.2 Nonlinear Dynamic Inversion Control

In the absence of disturbances d_1, d_2 and d_3, an autopilot for the missile to track an angle-of-attack reference $\omega(t)$ may be designed using the nonlinear dynamic inversion control (NDIC) [117]. The output is chosen as

$$y = \alpha + k_q q, \tag{7.30}$$

where k_q is a chosen constant. The resultant control law is designed by

$$u_{ndic} = \delta - [t_1/(b_1 + k_q b_2)]\{k_1(y - \omega) + k_2[f_1 + q + b_1 \delta \\ + k_q(f_2 + b_2 \delta) - \dot{\omega}] + m - \ddot{\omega}\}, \tag{7.31}$$

where

$$m = m_1[f_1(\alpha) + q + b_1(\alpha)\delta] + f_2(\alpha) + b_2\delta, \tag{7.32}$$

$$m_1 = \frac{\partial f_1(\alpha)}{\partial \alpha} + \frac{\partial b_1(\alpha)}{\partial \alpha}\delta + k_q \frac{\partial f_2(\alpha)}{\partial \alpha}, \tag{7.33}$$

and k_1, k_2 are constant gains to be designed according to the desired closed-loop behaviors.

Assume that the command signal ω_{cmd} is filtered by a low-pass prefilter so as to provide the reference for tracking

$$G(s) = \omega_n^2/(s^2 + 2\varsigma\omega_n s + \omega_n^2). \tag{7.34}$$

Substituting the NDIC law (7.31) into the longitudinal dynamics of missile, the closed-loop error dynamics are given by

$$\ddot{y}(t) - \ddot{\omega}(t) + k_2[\dot{y}(t) - \dot{\omega}(t)] + k_1[y(t) - \omega(t)] = 0. \tag{7.35}$$

The parameters in Equations (7.30)–(7.35) are chosen as

$$\varsigma = 1.0, \quad \omega_n = 6(\text{rad/s}), \tag{7.36}$$

$$k_q = 0.06(\text{s}), \tag{7.37}$$

$$k_1 = 15(1/\text{s}^2), \quad k_2 = 6(1/\text{s}). \tag{7.38}$$

As shown in (7.35), the longitudinal dynamics of missile are feedback linearized by NDIC. The closed-loop poles under NDIC is given by $-7.0 \pm 7.14j$, which implies that promising tracking performance is achieved under the control law in the absence of disturbances. However, it is well known that such NDIC scheme has poor robustness and disturbance rejection ability [40, 118].

7.4.3 Nonlinear Disturbance Observer-Based Robust Control

The state vector of missile system is defined as $x = [\alpha, q, \delta]^T$. Combining Equations (7.23), (7.24) with (7.29) gives

$$
\underbrace{\begin{bmatrix} \dot{\alpha} \\ \dot{q} \\ \dot{\delta} \end{bmatrix}}_{\dot{x}} = \underbrace{\begin{bmatrix} f_1(\alpha) + q + b_1(\alpha)\delta \\ f_2(\alpha) + b_2\delta \\ -(1/t_1)\delta \end{bmatrix}}_{f(x)} + \underbrace{\begin{bmatrix} 0 \\ 0 \\ 1/t_1 \end{bmatrix} u}_{g_1(x)\,u} + \underbrace{\begin{bmatrix} 1 & 0 & 0 \\ 0 & 1 & 0 \\ 0 & 0 & 1 \end{bmatrix} \begin{bmatrix} d_1 \\ d_2 \\ d_3 \end{bmatrix}}_{g_2(x)\,d} \quad (7.39)
$$

Output Equation (7.30) is rewritten as

$$
y = h(x) = \alpha + k_q q. \tag{7.40}
$$

It can be observed from Equations (7.39) and (7.40) that the missile system has the same formulation as system (7.1).

Using nonlinear disturbance observer (7.2), the lumped disturbances in the missile system can be estimated. To determine the nonlinear disturbance compensation gain for the missile system following from the procedure proposed in Section 7.3, the above dynamics are reformulated as (7.17). Nonlinear matrix-value functions $g_1(x)$ and $g_2(x)$ are the same as those in (7.39), while $\bar{f}(x)$ and $\bar{h}(x)$ are expressed as follows

$$
\bar{f}(x) = \begin{bmatrix} f_1'(\alpha) & 1 & b_1(\alpha) \\ f_2'(\alpha) & 0 & b_2 \\ 0 & 0 & -1/t_1 \end{bmatrix}, \tag{7.41}
$$

$$
\bar{h}(x) = \begin{bmatrix} 1, & k_q, & 0 \end{bmatrix}, \tag{7.42}
$$

where

$$
f_1'(\alpha) = \frac{180g\,QS}{\pi\,WV}\cos(\frac{\pi\alpha}{180})(1.03 \times 10^{-4}\alpha^2 - 9.45 \times 10^{-3}|\alpha|
$$

$$
- 1.7 \times 10^{-1}), \tag{7.43}
$$

$$
f_2'(\alpha) = \frac{180\,QSd}{\pi\,I_{yy}}(2.15 \times 10^{-4}\alpha^2 - 1.95 \times 10^{-2}|\alpha| + 5.1 \times 10^{-2}). \tag{7.44}
$$

Rearranging NDIC law (7.31) yields

$$
u_{ndic} = \bar{\alpha}(x)x + \gamma(\omega, \dot{\omega}, \ddot{\omega}), \tag{7.45}
$$

where $\bar{\alpha}(x) = [\bar{\alpha}_1(x), \bar{\alpha}_2(x), \bar{\alpha}_3(x)]$, and

$$\bar{\alpha}_1(x) = -\frac{t_1}{b_1 + k_q b_2} \left[k_1 + (k_2 + m_1) f_1'(\alpha) + (1 + k_2 k_q) f_2'(\alpha) \right], \quad (7.46)$$

$$\bar{\alpha}_2(x) = -\frac{t_1}{b_1 + k_q b_2} (k_1 k_q + k_2 + m_1), \quad (7.47)$$

$$\bar{\alpha}_3(x) = 1 - \frac{t_1}{b_1 + k_q b_2} (k_2 b_1 + k_2 k_q b_2 + m_1 b_1 + b_2), \quad (7.48)$$

$$\gamma(\omega, \dot{\omega}, \ddot{\omega}) = \frac{t_1}{b_1 + k_q b_2} (k_1 \omega + k_2 \dot{\omega} + \ddot{\omega}). \quad (7.49)$$

Since all nonlinear functions including $\bar{f}(x)$, $\bar{h}(x)$, $g_1(x)$, $g_2(x)$, and $\bar{\alpha}(x)$ are available, nonlinear disturbance compensation gain $\beta(x)$ is obtained by using (7.19). The composite nonlinear disturbance observer-based robust control (NDOBRC) law is then given by

$$u = u_{ndic} + \beta(x)\hat{d}, \quad (7.50)$$

where u_{ndic} is NDIC law (7.31), $\beta(x)$ is the disturbance compensation vector given by (7.19), and $\hat{d}$ is disturbance estimate governed by (7.2). The control structure of the proposed NDOBRC scheme is shown in Figure 7.1.

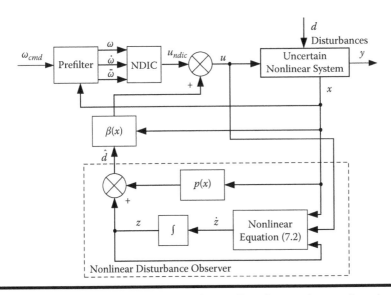

Figure 7.1 **Control structure of the nonlinear disturbance observer-based robust control (NDOBRC) scheme.**

7.4.4 Simulation Studies

The nonlinear vector-valued function $l(x)$ in NDO (7.2) is chosen as

$$l(x) = \begin{bmatrix} 10 & 0 & 0 \\ 0 & 10 & 0 \\ 0 & 0 & 10 \end{bmatrix}. \tag{7.51}$$

It is well known that the NDIC method has a poor disturbance rejection ability and robustness [40]. Integral action provides a practical way to eliminate the steady-state error in the presence of disturbances/uncertainties [42]. To demonstrate the effectiveness of the proposed method, in addition to the NDIC method, NDIC plus an integral action (called NDIC+I) is also employed for performance comparison. The control law of NDIC+I is given as follows

$$
\begin{aligned}
u_{ndic+i} &= u_{ndic} - [t_1/(b_1 + k_q b_2)] k_0 \int (y - w) d\tau \\
&= \delta - [t_1/(b_1 + k_q b_2)] \{ k_0 \int (y - w) d\tau + k_1 (y - w) \\
&\quad + k_2 [f_1 + q + b_1 \delta + k_q (f_2 + b_2 \delta) - \dot{w}] + m - \ddot{w} \},
\end{aligned} \tag{7.52}
$$

where u_{ndic} is the NDIC law in (7.31), and the integral coefficient is selected as

$k_0 = 20$ so as to achieve a satisfactory performance.

It can be seen from control law (7.52) that the NDIC+I method has a PID like structure. In addition, the closed-loop system under (7.52) is given by

$$\ddot{y} - \ddot{w} + k_2(\dot{y} - \dot{w}) + k_1(y - w) + k_0 \int (y - w) d\tau = 0. \tag{7.53}$$

It will be shown by simulation later that the offset caused by disturbances and uncertainties can be eliminated by the NDIC+I method.

7.4.4.1 External Disturbance Rejection Ability

The external disturbance rejection ability of missile system under the proposed NDOBRC method is investigated in this subsection. Suppose that the external disturbance $d_2 = 5$ deg/s² enters at $t = 2$ sec. The response curves of both the output and input under three control methods are shown in Figure 7.2. The corresponding response curves of the states are shown in Figure 7.3.

It can be observed from Figure 7.2(a) that the NDIC method results in a large offset. As for the NDIC+I method, there is no offset any more, but large overshoot

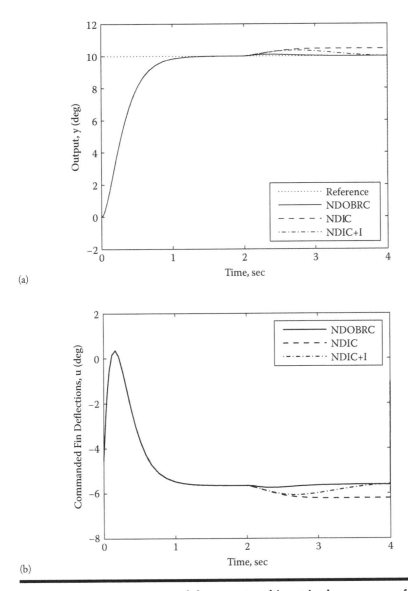

(a)

(b)

Figure 7.2 **Response curves of the output and input in the presence of external disturbances under the control laws of NDOBRC (7.50) (solid line), NDIC (7.31) (dashed line), and NDIC+I (7.52) (dash-dotted line). The reference signal is denoted by dotted line.**

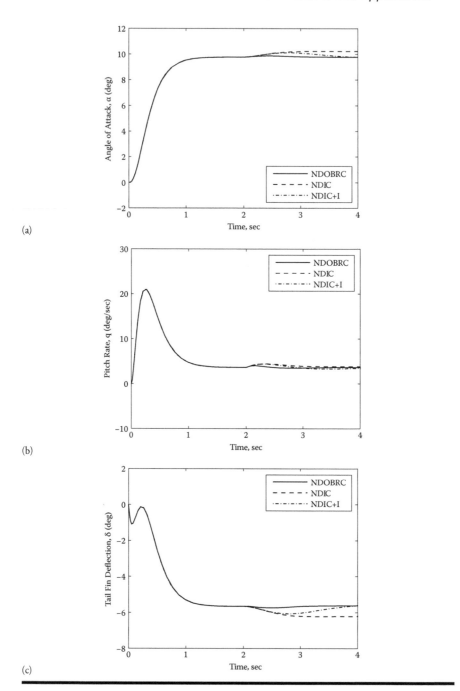

(a)

(b)

(c)

Figure 7.3 Response curves of the states in the presence of external disturbances under the control laws of NDOBRC (7.50) (solid line), NDIC (7.31) (dashed line), and NDIC+I (7.52) (dash-dotted line).

and long settling time are experienced. The proposed NDOBRC provides a much better transient and steady-state performance, such as small overshoot, short settling time and offset free property. The control input in Figure 7.2(b) shows that no high gain and excessive energy are demanded for all the three methods.

It is shown by Figure 7.3 that all states under the three methods remain within allowable regions. This implies that the proposed method obtains much better external disturbance rejection performance than the other two methods without bringing any adverse effects to all the states.

7.4.4.2 Robustness Against Model Uncertainties

The robustness against model uncertainties of the proposed NDOBRC is tested in this subsection. Two cases of model uncertainties are considered to investigate the performance of robustness in detail.

Case I: both $f_1(\alpha)$ and $f_2(\alpha)$ have variations of +20%.

For the first case of model uncertainties, response curves of the output/input and the states under the three control methods are shown by Figures 7.4 and 7.5, respectively. As shown by Figure 7.4(a), the NDIC approach has resulted in large offset. For the NDIC+I method, the offset is eliminated but quite slowly. The proposed NDOBRC method tracks the reference command rapidly without any offset.

It is shown by Figures 7.4(b) and 7.5 that both the control input and the states remain within allowable regions. These variables under the NDOBRC method converge to the desired equilibrium point much quicker than the other two methods.

Case II: $f_1(\alpha)$ and $f_2(\alpha)$ have variations of -20% and -7%, respectively.

In such a case of uncertainties, Figures 7.6 and 7.7 show response curves of the output/input and states under the three control methods, respectively. It can be shown by Figure 7.6(a) that the output under the NDIC method substantially departs from the desired reference trajectory. This implies that the missile becomes essentially unstable as the model is only valid when $|\alpha| \leq 20$ degrees, but the angle of attack reaches 70 degrees in simulation. Figure 7.6(a) shows that the output under the NDIC+I method becomes oscillating and unstable. The proposed NDOBRC has achieved the best performance, including a small overshoot, a short settling time, no oscillation and offset free.

It is shown by Figure 7.6(b) that the magnitude of control input under the NDOBRC is much smaller than those under the other two methods. The NDIC method is unusable in this case since the magnitude of control input is huge and over the actuator constraint. Also from the response curves of states in Figure 7.7, it can be concluded that the proposed method achieves the best performance of robustness among all the three methods.

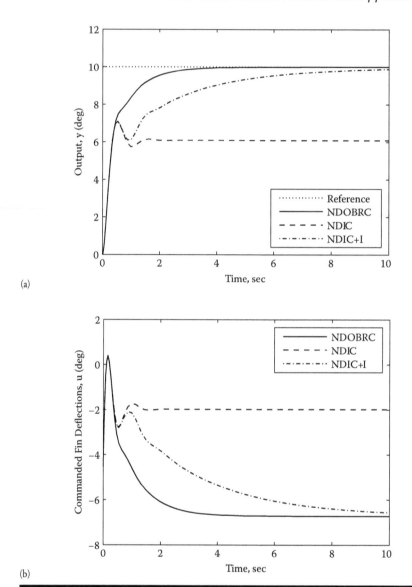

(a)

(b)

Figure 7.4 **Response curves of the output and input in the first case of model uncertainties (both $f_1(\alpha)$ and $f_2(\alpha)$ have variations of +20%) under the control laws of NDOBRC (7.50) (solid line), NDIC (7.31) (dashed line), and NDIC+I (7.52) (dash-dotted line). The reference signal is denoted by dotted line.**

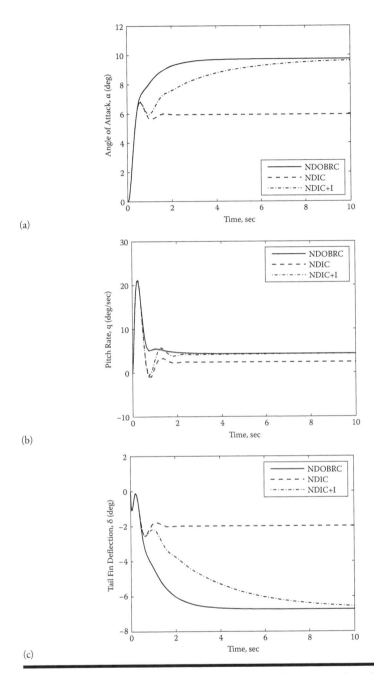

(a)

(b)

(c)

Figure 7.5 **Response curves of the states in the first case of model uncertainties (both $f_1(\alpha)$ and $f_2(\alpha)$ have variations of +20%) under the control laws of NDO-BRC (7.50) (solid line), NDIC (7.31) (dashed line), and NDIC+I (7.52) (dash-dotted line).**

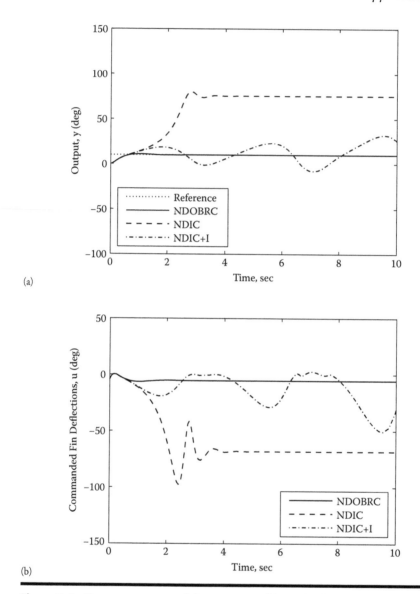

(a)

(b)

Figure 7.6 **Response curves of the output and input in the second case of model uncertainties ($f_1(\alpha)$ and $f_2(\alpha)$ have variations of -20% and -7%, respectively) under the control laws of NDOBRC (7.50) (solid line), NDIC (7.31) (dashed line), and NDIC+I (7.52) (dash-dotted line). The reference signal is denoted by dotted line.**

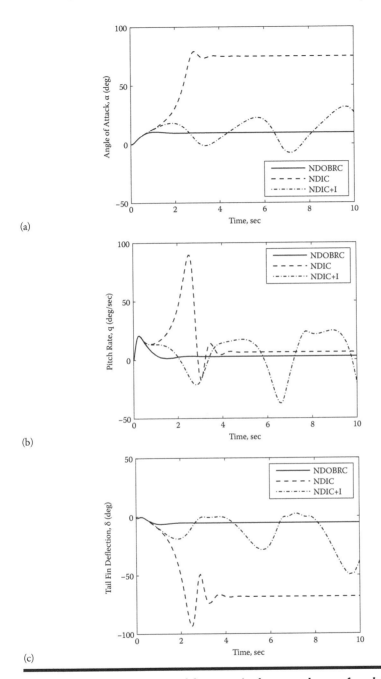

(a)

(b)

(c)

Figure 7.7 Response curves of the states in the second case of model uncertainties ($f_1(\alpha)$ and $f_2(\alpha)$ have variations of -20% and -7%, respectively) under the control laws of NDOBRC (7.50) (solid line), NDIC (7.31) (dashed line), and NDIC+I (7.52) (dash-dotted line).

7.5 Summary

As clearly demonstrated in the missile example in this chapter, external disturbances and parameter perturbations always bring adverse effects to stability and control performance modern industrial systems. DOBC has provided a solution to this problem. It can significantly improve disturbance attenuation ability and robustness against uncertainties, and acts like a "patch" to the existing feedback control design without considerably changing the nominal control design. The existing DOBC methods are only applicable to matched disturbances. A NDOBRC method has been proposed for nonlinear systems in the presence of mismatched disturbances and uncertainties. It has been shown that by appropriately designing the nonlinear compensation gains, offset free tracking error can be achieved on system output. Simulation studies of a missile system have been carried out to show the validity of the proposed NDOBRC method. The results have shown that the proposed method achieves much better disturbance rejection ability and robustness against model uncertainties as compared with NDIC and NDIC+I methods.

Chapter 8

Nonlinear Disturbance Observer-Based Control for Systems with Arbitrary Disturbance Relative Degrees

8.1 Introduction

The nonlinear disturbance observer-based control (NDOBC) is proposed and further researched by Chen and his colleagues [10, 2, 40, 5, 86]. It should be pointed out that these results are only valid for the case where the disturbance relative degree (DRD) is higher than or equal to the input relative degree (IRD) [10]. Disturbance attenuation of nonlinear system with the DRD strictly lower than the IRD purely via a NDOBC without resorting to a high-gain feedback-based design is a longstanding unresolved problem. In fact, the condition that the DRD is lower than the IRD implies that the so-called matching condition [100] for disturbance is not satisfied anymore. As for the mismatched disturbance, it is impossible to transfer the disturbances to the same channel as the control inputs by any coordinate transformation [106], which brings barriers to the control design of such systems.

A novel NDOBC is proposed in this chapter for nonlinear systems with arbitrary DRD via feedback linearization and disturbance observer-based feedforward compensation. A systematic approach is proposed for the design of disturbance compensation gain. It is shown that the offset of closed-loop system with disturbance can be eliminated in steady-state from the output channel under the proposed control method. This chapter mainly refers to [16, 17].

8.2 Problem Formulation

Consider an affine nonlinear system with the form

$$\begin{cases} \dot{x} = f(x) + g(x)u + p(x)d \\ y = h(x) \end{cases} \tag{8.1}$$

where $x \in R^n$, $u \in R$, $d \in R$, and $y \in R$ are the state vector, the control input, the unknown disturbances and the output, respectively. It is supposed that $f(x)$, $g(x)$, and $p(x)$ are smooth functions in terms of x.

To address the problem here, the standard Lie derivative notation is utilized in this chapter [106].

Definition 8.1 [106] *The relative degree from the control input to the output of system (8.1) is $\rho \leq n$, which implies that $L_g L_f^k h(x) = 0$ for all $k < \rho - 1$ and $L_g L_f^{\rho-1} h(x) \neq 0$ for all $x \in R^n$. For simplicity, it is referred to as the input relative degree (IRD). Similarly, the disturbance relative degree (DRD) can be defined as v.*

Remark 8.1 [10, 106] *If the DRD of system (8.1) is higher than the IRD, that is, $v > \rho$, the disturbance attenuation problem of system (8.1) can be solved by exact disturbance decoupling and no disturbance-based feedforward compensation is required. Actually, the state feedback control*

$$u = \alpha(x) + \beta(x)v \tag{8.2}$$

with

$$\alpha(x) = -\frac{L_f^\rho h(x)}{L_g L_f^{\rho-1} h(x)} \tag{8.3}$$

and

$$\beta(x) = \frac{1}{L_g L_f^{\rho-1} h(x)} \tag{8.4}$$

solves the problem.

Remark 8.2 [10, 106] *If the DRD of system (8.1) equals to the IRD, that is, $v = \rho$, the disturbance attenuation problem of system (8.1) can be solved by the following disturbance based feedforward control*

$$u = \alpha(x) + \beta(x)v + \gamma(x)d \tag{8.5}$$

where $\alpha(x)$ and $\beta(x)$ are the same as those in (8.3) and (8.4), and $\gamma(x)$ is designed as

$$\gamma(x) = -\frac{L_p L_f^{\rho-1} h(x)}{L_g L_f^{\rho-1} h(x)}. \tag{8.6}$$

Remark 8.3 *Most DOBC approaches are concerned with the case where the disturbance acts on the same channel as that of control input, that is, $g(x) = p(x)$. This is a special case of the result in Remark 8.2 since it can be obtained from (8.6) under the condition $g(x) = p(x)$ that $\gamma(x) = -1$, which is the result of most DOBC methods, e.g., see* [2, 40, 5, 86].

When it comes to the case that the DRD is strictly lower than the IRD, that is, $\nu < \rho$, the existing DOBC methods are not available anymore. In this case, the disturbances inevitably affect the states no matter which feedback-based method is used. The most achievable design goal in this case would be to find a control method such that the disturbances do not affect the interested output (at least in steady-state), which motivates further research on NDOBC for nonlinear system with arbitrary DRD.

8.3 NDOBC for SISO Nonlinear System with Arbitrary DRD

For system (8.1), it is always possible to choose a group of coordinate transformation, $z_i = \phi_i(x)$, $i = 1, 2, \ldots, n$ to feedback linearize nonlinear system (8.1) in the absence of disturbance, where $z_i = \phi_i(x) = L_f^{i-1} h(x)$, $i = 1, 2, \ldots, \rho$ and $z_{\rho+1} = \phi_{\rho+1}(x), \ldots, z_n = \phi_n(x)$ satisfying $L_g \phi_i(x) = 0$ for all $\rho + 1 \leq i \leq n$ such that the mapping

$$\Phi(x) = \begin{bmatrix} \phi_1(x) \\ \vdots \\ \phi_n(x) \end{bmatrix}$$

has a nonsingular Jacobian matrix [106].

The description of system (8.1) under the new coordinates $z_i = \phi_i(x), 1 \leq i \leq n$ is then expressed as

$$\frac{dz_1}{dt} = L_f h(x) + L_p h(x) d \tag{8.7}$$

$$\vdots$$

$$\frac{dz_{\rho-1}}{dt} = L_f^{\rho-1} h(x) + L_p L_f^{\rho-2} h(x) d \tag{8.8}$$

For z_i, $i = \rho$, we obtain

$$\frac{dz_\rho}{dt} = L_f^\rho h(x) + L_g L_f^{\rho-1} h(x) u + L_p L_f^{\rho-1} h(x) d \tag{8.9}$$

For z_i, $\rho + 1 \le i \le n$, we have

$$\frac{dz_i}{dt} = L_f \phi_i(x) + L_p \phi_i(x) d \tag{8.10}$$

Replace $x(t)$ with its expression as a function of $z(t)$, that is, $x(t) = \Phi^{-1}(z(t))$, and let

$$a(z) = L_g L_f^{\rho-1} h(\Phi^{-1}(z)) \tag{8.11}$$

$$b(z) = L_f^\rho h(\Phi^{-1}(z)) \tag{8.12}$$

$$m_i(z) = L_p L_f^{i-1} h(\Phi^{-1}(z)), \, i = 1, \dots, \rho \tag{8.13}$$

$$q(z) = \left(q_1(z), \, \dots, \, q_{n-\rho}(z) \right)^T \tag{8.14}$$

$$q_i(z) = L_f \phi_{\rho+i}(\Phi^{-1}(z)), \, i = 1, \dots, n - \rho \tag{8.15}$$

$$r(z) = \left(r_1(z), \, \dots, \, r_{n-\rho}(z) \right)^T \tag{8.16}$$

$$r_i(z) = L_p \phi_{\rho+i}(\Phi^{-1}(z)), \, i = 1, \dots, n - \rho \tag{8.17}$$

Rearranging (8.7)–(8.10) and considering notations (8.11)–(8.17), the state-space description of nonlinear system under the new coordinates is obtained as

$$\begin{cases} \dot{z}_1 = z_2 + m_1(z)d \\ \dot{z}_2 = z_3 + m_2(z)d \\ \quad \vdots \\ \dot{z}_{\rho-1} = z_\rho + m_{\rho-1}(z)d \\ \dot{z}_\rho = b(z) + a(z)u + m_\rho(z)d \\ \dot{z}_{\rho+1} = q_1(z) + r_1(z)d \\ \quad \vdots \\ \dot{z}_n = q_{n-\rho}(z) + r_{n-\rho}(z)d \\ y = z_1 \end{cases} \tag{8.18}$$

8.3.1 Control Law Design

The disturbance attenuation problem of nonlinear system with arbitrary DRD via a novel NDOBC is presented via the following theorem.

Theorem 8.1 *The disturbance of nonlinear system (8.1) can be removed from the output channel in steady-state by the following composite control law*

$$u = \alpha(x) + \beta(x)v + \gamma(x)\hat{d} \qquad (8.19)$$

with $\alpha(x)$ and $\beta(x)$ being the same as those in (8.3) and (8.4), $\hat{d}$ is the disturbance estimation by the NDO in (7.2),

$$\gamma(x) = -\frac{\sum_{i=1}^{\rho-1} c_i L_p L_f^{i-1} h(x) + L_p L_f^{\rho-1} h(x)}{L_g L_f^{\rho-1} h(x)} \qquad (8.20)$$

and

$$v = -\sum_{i=0}^{\rho-1} c_i L_f^i h(x) \qquad (8.21)$$

where parameters c_i $(i = 0, 1, \ldots, \rho - 1)$ have to be designed such that the polynomial

$$p_0(s) = c_0 + c_1 s + \cdots + c_{\rho-1} s^{\rho-1} + s^\rho \qquad (8.22)$$

is Hurwitz stable.

Proof Considering $x(t) = \Phi^{-1}(z(t))$ and the notations in (8.11)-(8.13), composite control law (8.19) can be rewritten as

$$u = -\frac{b(z) + \sum_{i=0}^{\rho-1} c_i z_{i+1} + \sum_{i=1}^{\rho-1} c_i m_i(z)\hat{w} + m_\rho(z)\hat{d}}{a(z)} \qquad (8.23)$$

Let $\xi = (z_1, \ldots, z_\rho)^T$, $\eta = (z_{\rho+1}, \ldots, z_n)^T$, and $C = (1, 0, \ldots, 0)_{1\times\rho}$. Considering disturbance estimation error system (7.4) and substituting (8.23) into (8.18) yields

$$\begin{cases} \dot{\xi} = A\xi + M(\xi, \eta)d + N(\xi, \eta)e \\ \dot{\eta} = q(\xi, \eta) + r(\xi, \eta)d \\ y = C\xi \end{cases} \qquad (8.24)$$

where

$$A = \begin{bmatrix} 0 & 1 & 0 & \cdots & 0 \\ 0 & 0 & 1 & \cdots & 0 \\ \cdot & \cdot & \cdot & \cdots & \cdot \\ 0 & 0 & 0 & \cdots & 1 \\ -c_0 & -c_1 & -c_2 & \cdots & -c_{p-1} \end{bmatrix},$$

$$M(\xi, \eta) = \begin{bmatrix} m_1(\xi, \eta) \\ m_2(\xi, \eta) \\ \vdots \\ m_{p-1}(\xi, \eta) \\ -\sum_{i=1}^{p-1} c_i m_i(\xi, \eta) \end{bmatrix},$$

$$N(\xi, \eta) = \begin{bmatrix} 0 \\ 0 \\ \vdots \\ 0 \\ m_p(\xi, \eta) + \sum_{i=1}^{p-1} c_i m_i(\xi, \eta) \end{bmatrix}$$

It can be obtained from (8.24) that

$$\xi = A^{-1} \left[\dot{\xi} - M(\xi, \eta)d - N(\xi, \eta)e \right] \quad (8.25)$$

Suppose that the closed-loop system reaches a steady-state and let ξ_s, η_s, e_s, and y_s denote the steady-state values of ξ, η, e, and y, respectively. Steady-state value ξ_s can be represented as

$$\xi_s = -A^{-1}[M(\xi_s, \eta_s)d_s + N(\xi_s, \eta_s)e_s] \quad (8.26)$$

Considering the fact that $e_s = 0$ and combining (8.26) with (8.24) gives

$$y_s = -CA^{-1}M(\xi_s, \eta_s)d_s$$

$$= -[1, 0, \cdots, 0] \times \begin{bmatrix} -\frac{c_1}{c_0} & -\frac{c_2}{c_0} & \cdots & -\frac{c_{p-1}}{c_0} & -\frac{1}{c_0} \\ 1 & 0 & \cdots & 0 & 0 \\ 0 & 1 & \cdots & 0 & 0 \\ \cdot & \cdot & \cdots & \cdot & \cdot \\ 0 & 0 & \cdots & 1 & 0 \end{bmatrix}$$

$$\times \begin{bmatrix} m_1(\xi_s, \eta_s) \\ m_2(\xi_s, \eta_s) \\ \vdots \\ m_{\rho-1}(\xi_s, \eta_s) \\ -\sum_{i=1}^{\rho-1} c_i m_i(\xi_s, \eta_s) \end{bmatrix} d_s$$

$$= 0. \tag{8.27}$$

Remark 8.4 *The proposed method is a generalized case of the existing NDOBC methods in the previous literature. In fact, $v = \rho$ implies that $L_p L_f^i h(x) = 0$ for $i = 0, 1, \ldots, \rho - 2$. Considering $p(x) = g(x)$ and these conditions, one obtains from (8.20) that $\gamma(x) = -1$.*

8.3.2 Stability Analysis

Assumption 8.1 *The zero dynamics of nonlinear system (8.1) are stable, which implies the nonlinear system is of minimum phase.*

In the absence of disturbance, that is, $d = 0$, the closed-loop system can be formulated from (8.24) and expressed as

$$\begin{cases} \dot{\xi} = A\xi \\ \dot{\eta} = q(\xi, \eta) \end{cases} \tag{8.28}$$

Lemma 8.1 *[106] Suppose the equilibrium $\eta = 0$ of zero dynamics $\dot{\eta} = q(0, \eta)$ of system (8.28) are locally asymptotically stable and all the roots of polynomial $p_0(s)$ in (8.22) have negative real parts. Then the feedback control law $u = \alpha(x) + \beta(x)v$ with $\alpha(x), \beta(x),$ and v designed as (8.3), (8.4) and (8.21), locally asymptotically stabilizes system $\dot{x} = f(x) + g(x)u$ in the equilibrium $(\xi, \eta) = (0, 0)$.*

Lemma 8.2 *[119] Consider the following cascade-connected system*

$$\begin{cases} \dot{x} = \overline{F}(x, e) \\ \dot{e} = \overline{G}(e) \end{cases} \tag{8.29}$$

Suppose the equilibrium $x = 0$ of $\dot{x} = \overline{F}(x, 0)$ is locally asymptotically stable and the equilibrium $e = 0$ of $\dot{e} = \overline{G}(e)$ is locally asymptotically stable. Then, the equilibrium $(x, e) = (0, 0)$ is locally asymptotically stable.

Lemma 8.3 [87] *Consider system $\dot{x} = \underline{F}(X, w)$ with $F : D \times D_w \rightarrow R^n$ is locally Lipschitz in X and w, $D \subset R^n$ is a domain that contains $X = 0$, and $D_w \subset R^m$ is a domain that contains $w = 0$. If $F(X, w)$ is continuously differentiable and the origin of system $\dot{x} = \underline{F}(X, 0)$ is asymptotically stable, then $\dot{x} = \underline{F}(X, w)$ is locally ISS.*

The input-to-state stability of nonlinear system (8.1) under composite control law $u = \alpha(x) + \beta(x)v + \gamma(x)\hat{d}$ is established by the following theorem.

Theorem 8.2 *Nonlinear system (8.1) under composite control law (8.19) is locally ISS if the following conditions are satisfied:*

(i) *Assumption 7.1 is satisfied, that is, the zero dynamics of nonlinear system $\dot{x} = f(x) + g(x)u$ under the control of $u = \alpha(x) + \beta(x)v$ are stable.*

(ii) *All the roots of the polynomial $p_0(s)$ in (8.22) have negative real parts,*

(iii) *The observer gain $l(x)$ in (7.2) is selected such that system (7.4) is asymptotically stable,*

(iv) *The disturbance compensation gain is selected such that $F(x, e, w) = f(x) + g(x)[\alpha(x)+\beta(x)v]+g(x)\gamma(x)e+[g(x)\gamma(x)+p(x)]w$ and $\overline{G}(e) = -l(x)p(x)$ e are continuously differentiable.*

Proof Substituting the control law into system (8.1), the augmented closed-loop system is given by

$$\begin{cases} \dot{x} = F(x, e, w) \\ \dot{e} = \overline{G}(e) \end{cases} \tag{8.30}$$

where

$$F(x, e, w) = f(x) + g(x)[\alpha(x) + \beta(x)v] + g(x)\gamma(x)e + [g(x)\gamma(x) + p(x)]w$$

and

$$\overline{G}(e) = -l(x)p(x)e$$

Since conditions (i) and (ii) are satisfied, it can be obtained from Lemma 8.1 that system $\dot{x} = F(x, 0, 0)$ is locally asymptotically stable in the equilibrium $(\xi, \eta) = (0, 0)$.

Let $\overline{F}(x, e) = F(x, e, 0)$, then one obtains from the above result that $\dot{x} = \overline{F}(x, 0)$ is locally asymptotically stable in the equilibrium $x = 0$. Combining this conclusion with condition (iii), it follows from Lemma 8.2 that system

$$\begin{cases} \dot{x} = \overline{F}(x, e) = F(x, e, 0) \\ \dot{e} = \overline{G}(e) = -l(x)p(x)e \end{cases} \tag{8.31}$$

is locally asymptotically stable in the equilibrium $(x, e) = (0, 0)$.

Let $X = [x, e]^T$ and $\underline{F}(X, d) = \left[F(x, e, d), \overline{G}(e) \right]^T$, then systems (8.30) and (8.31) can be denoted as

$$\dot{X} = \underline{F}(X, d) \tag{8.32}$$

and

$$\dot{X} = \underline{F}(X, 0) \tag{8.33}$$

Condition (iv) guarantees that $\underline{F}(X, w)$ is continuously differentiable. Since system (8.33) is locally asymptotically stable, it can be followed from Lemma 8.3 that system (8.32) is locally input-to-state stable, which completes the proof of this theorem.

8.4 NDOBC for MIMO Nonlinear Systems with Arbitrary DRDs

Consider a MIMO nonlinear system depicted by

$$\begin{cases} \dot{x} = f(x) + g(x)u + p(x)d, \\ y = h(x), \end{cases} \tag{8.34}$$

where $x = [x_1, \ldots, x_n]^T \in R^n$, $u = [u_1, \ldots, u_m]^T \in R^m$, $d = [d_1, \ldots, d_n]^T \in R^n$ and $y = [y_1, \ldots, y_m]^T \in R^m$ denote the state, input, disturbance and output vectors, respectively. $f(x), g(x) = [g_1(x), \ldots, g_m(x)]$, $p(x) = [p_1(x), \ldots, p_n(x)]$, and $h(x) = [h_1(x), \ldots, h_m(x)]^T$ are smooth vector or matrix fields on R^n. Without loss of generality, it is supposed that the equilibrium of system (8.34) in the absence of disturbances is $x_0 = 0$. The standard Lie derivative notation is used in this book, stated as follows [106].

Definition 8.2 [106] *The vector relative degree from the control inputs to the outputs of system (8.34) is $(\sigma_1, \ldots, \sigma_m)$ at the equilibrium x_0 if $L_{g_j} L_f^k h_i(x) = 0$ $(1 \leq j \leq m, 1 \leq i \leq m)$ for all $k < \sigma_i - 1$, and for all x in a neighborhood of x_0, and the $m \times m$ matrix*

$$A(x) = \begin{bmatrix} L_{g_1} L_f^{\sigma_1 - 1} h_1 & L_{g_2} L_f^{\sigma_1 - 1} h_1 & \cdots & L_{g_m} L_f^{\sigma_1 - 1} h_1 \\ L_{g_1} L_f^{\sigma_2 - 1} h_2 & L_{g_2} L_f^{\sigma_2 - 1} h_2 & \cdots & L_{g_m} L_f^{\sigma_2 - 1} h_2 \\ \vdots & \vdots & \ddots & \vdots \\ L_{g_1} L_f^{\sigma_m - 1} h_m & L_{g_2} L_f^{\sigma_m - 1} h_m & \cdots & L_{g_m} L_f^{\sigma_m - 1} h_m \end{bmatrix}, \tag{8.35}$$

is nonsingular at $x = x_0$. For simplicity, it is referred to as the input relative degree (IRD). Similarly, the disturbance relative degree (DRD) at x_0 is defined as $(v_1, \ldots, v_m)$.

Remark 8.5 [10] *The existing nonlinear disturbance observer based control (NDOBC) is only available for nonlinear system (8.34) whose DRDs are higher than or equal to its IRDs, that is, $v_i \geq \sigma_j$ for all $1 \leq i \leq m$ and $1 \leq j \leq m$.*

Remark 8.6 *When it comes to the case that some DRD of nonlinear system (8.34) is strictly lower than some IRD, that is, $v_i < \sigma_j$ for some $i, j \in \{1, \ldots, m\}$, the disturbances will inevitably affect the states regardless of which feedback control method is employed. In this case, the most sensible design goal would be to find a control method such that the disturbances do not affect the interested output variables (at least in steady-state). As pointed in Remark 8.1, the existing NDOBC is no longer available for this case, which motivates further research on generalized NDOBC for system (8.34) with arbitrary DRDs.*

The nonlinear disturbance observer (NDO) proposed by [2] still provides an adequate way to estimate the disturbances in system (8.34), given by

$$\begin{cases} \dot{z} = -l(x)[p(x)(\lambda(x) + z) + f(x) + g(x)u], \\ \hat{d} = z + \lambda(x), \end{cases} \tag{8.36}$$

where $\hat{d} = [\hat{d}_1, \ldots, \hat{d}_n]^T$ and z are the estimation vector of disturbance vector d and the internal state vector of nonlinear observer, respectively. $\lambda(x)$ is a nonlinear function to be designed. Observer gain $l(x)$ is designed as

$$l(x) = \frac{\partial \lambda(x)}{\partial x}.$$

The disturbance estimation error of (8.36) for system (8.34) is obtained by

$$\dot{e}(t) = -l(x)p(x)e(t) + \dot{d}, \tag{8.37}$$

where the estimation error is defined as $e = d - \hat{d}$.

Assumption 8.2 *The derivatives of disturbances in system (8.34) are bounded, i.e., $\|\dot{d}(t)\| < \infty$.*

This is a general assumption made for the disturbance estimation.

Lemma 8.4 [111] *Suppose that Assumption 8.2 is satisfied. The disturbance estimation error system (8.37) is locally input-to-state stable (ISS) if the observer gain $l(x)$ is chosen such that*

$$\dot{e}(t) + l(x)p(x)e(t) = 0, \tag{8.38}$$

is asymptotically stable.

Assumption 8.3 *The disturbances in (8.34) are bounded and satisfy* $\lim_{t \to \infty} \dot{d}(t) = 0$.

This assumption is made for steady-state analysis of disturbance estimation and compensation.

Lemma 8.5 *Suppose that Assumption 8.3 is satisfied. Disturbance estimation $\hat{d}$ of NDO (8.36) can asymptotically estimate disturbance w in system (8.34) if observer gain $l(x)$ is chosen such that (8.38) is asymptotically stable.*

The proof of this lemma can be easily derived by combining the result of ISS definition in [87].

8.4.1 Control Law Design

In order to present the main result, the disturbance decoupling for MIMO nonlinear systems with arbitrary DRDs is solved first.

Assumption 8.4 *The distribution $G = span\{g_1, \ldots, g_m\}$ of MIMO nonlinear system (8.34) is involutive.*

This is a necessary condition for state feedback linearization of MIMO nonlinear systems [106]. The disturbance decoupling problem for a MIMO nonlinear system with arbitrary DRDs is solved by the following theorem.

Theorem 8.3 *Consider a MIMO nonlinear system (8.34) with disturbances which has arbitrary DRDs satisfying the conditions in Assumptions 8.3 and 8.4. A disturbance decoupling control law, which can compensate the disturbances in the output channels of (8.34) in steady state, is given by*

$$u = A^{-1}(x)\left[-b(x) + v + \Gamma(x)d\right], \tag{8.39}$$

where $A(x)$ is the same as that in (8.35),

$$\Gamma(x) = \begin{bmatrix} \gamma_{11}(x) & \gamma_{12}(x) & \cdots & \gamma_{1n}(x) \\ \gamma_{21}(x) & \gamma_{22}(x) & \cdots & \gamma_{2n}(x) \\ \vdots & \vdots & \ddots & \vdots \\ \gamma_{m1}(x) & \gamma_{m2}(x) & \cdots & \gamma_{mn}(x) \end{bmatrix}, b(x) = \begin{bmatrix} b_1(x) \\ b_2(x) \\ \vdots \\ b_m(x) \end{bmatrix}, v = \begin{bmatrix} v_1 \\ v_2 \\ \vdots \\ v_m \end{bmatrix}$$

with

$$\gamma_{ij}(x) = -\sum_{k=0}^{\sigma_i-2} c_{k+1}^i L_{p_j} L_f^k h_i - L_{p_j} L_f^{\sigma_i-1} h_i, (i = 1, 2, \ldots, m; j = 1, 2, \ldots, n),$$

$$b_i(x) = L_f^{\sigma_i} h_i, v_i = -\sum_{k=0}^{\sigma_i-1} c_k^i L_f^k h_i, (i = 1, 2, \ldots, m),$$

where parameters $c_k^i (i = 1, 2, \ldots, m; k = 0, 1, \ldots, \sigma_i - 1)$ have to be designed such that the polynomials

$$p_0^i(s) = c_0^i + c_1^i s + \cdots + c_{\sigma_i-1}^i s^{\sigma_i-1} + s^{\sigma_i}, \tag{8.40}$$

are Hurwitz stable.

Proof The partial feedback linearization of nonlinear system (8.34) in the absence of disturbances is addressed firstly for the case of $\sigma = \sigma_1 + \sigma_2 + \ldots + \sigma_m < n$ since it is a general case (the case $\sigma = n$ is a special case and very straightforward)[106]. Under the condition of Assumption 8.4, a new group of coordinate transformations, which can feedback linearize system (8.34), is defined as [106]

$$\Phi(x) = \begin{bmatrix} \xi \\ \eta \end{bmatrix}, \tag{8.41}$$

where $\xi = \left[\xi^{1\,T}, \ldots, \xi^{m\,T} \right]^T$ with

$$\xi^i = \begin{bmatrix} \xi_1^i \\ \xi_2^i \\ \vdots \\ \xi_{\sigma_i}^i \end{bmatrix} = \begin{bmatrix} h_i(x) \\ L_f h_i(x) \\ \vdots \\ L_f^{\sigma_i-1} h_i(x) \end{bmatrix}, \tag{8.42}$$

for $1 \leq i \leq m$, and $\eta = [\eta_{\sigma+1}(x), \ldots, \eta_n(x)]^T$ is selected such that the mapping $\Phi(x)$ has a nonsingular Jacobian matrix and $L_{g_j}\eta_i(x) = 0$ for all $\sigma + 1 \leq i \leq n$ and $1 \leq j \leq m$.

The description of system (8.34) under the new coordinates $\Phi(x)$ is then expressed as $m + 1$ subsystems, and the ith one of the first m subsystems can be represented as

$$\Pi_i : \begin{cases} \dot{\xi}_1^i = \xi_2^i + \sum_{k=1}^n L_{p_k} h_i d_k, \\ \dot{\xi}_2^i = \xi_3^i + \sum_{k=1}^n L_{p_k} L_f h_i d_k, \\ \quad \vdots \\ \dot{\xi}_{\sigma_i}^i = L_f^{\sigma_i} h_i(x) + \sum_{k=1}^m L_{g_k} L_f^{\sigma_i-1} h_i u_k + \sum_{k=1}^n L_{p_k} L_f^{\sigma_i-1} h_i d_k, \\ y_i = \xi_1^i, \end{cases} \tag{8.43}$$

for all $1 \leq i \leq m$. The last subsystem has the following form

$$\Pi_{m+1} : \dot{\eta} = q(\xi, \eta) + r(\xi, \eta)d, \tag{8.44}$$

Let $\underline{\xi} = \left[\xi_{\sigma_1}^1, \ldots, \xi_{\sigma_m}^m\right]^T$, by collecting the last state equation in (8.43) for all $1 \leq i \leq m$ to formulate a new vector, it is obtained that

$$\dot{\underline{\xi}} = b(x) + A(x)u + D(x)d, \tag{8.45}$$

where

$$D(x) = \begin{bmatrix} L_{p_1} L_f^{\sigma_1-1} h_1 & L_{p_2} L_f^{\sigma_1-1} h_1 & \cdots & L_{p_n} L_f^{\sigma_1-1} h_1 \\ L_{p_1} L_f^{\sigma_2-1} h_2 & L_{p_2} L_f^{\sigma_2-1} h_2 & \cdots & L_{p_n} L_f^{\sigma_2-1} h_2 \\ \vdots & \vdots & \ddots & \vdots \\ L_{p_1} L_f^{\sigma_m-1} h_m & L_{p_2} L_f^{\sigma_m-1} h_m & \cdots & L_{p_n} L_f^{\sigma_m-1} h_m \end{bmatrix}.$$

Substituting control law (8.39) into (8.45), yields

$$\dot{\underline{\xi}} = v + \Gamma(x)w + D(x)d, \tag{8.46}$$

or an equivalent expression of

$$\dot{\xi}_{\sigma_i}^i = v_i + \sum_{k=1}^{n} (\gamma_{ik} + L_{p_k} L_f^{\sigma_i-1} h_i) d_k. \tag{8.47}$$

Combining (8.47) with (8.43), subsystem Π_i can be rewritten as

$$\Pi_i : \begin{cases} \dot{\xi}^i = \underline{A}^i \xi^i + \underline{D}^i(x)d, \\ y_i = C^i \xi^i, \end{cases} \tag{8.48}$$

for $1 \leq i \leq m$, where

$$\underline{A}^i = \begin{bmatrix} 0 & 1 & 0 & \cdots & 0 \\ 0 & 0 & 1 & \cdots & 0 \\ \vdots & \vdots & \vdots & \ddots & \vdots \\ 0 & 0 & 0 & \cdots & 1 \\ -c_0^i & -c_1^i & -c_2^i & \cdots & -c_{\sigma_i-1}^i \end{bmatrix},$$

$$\underline{D}^i(x) = \left[\underline{d}_1^i, \ldots, \underline{d}_n^i\right],$$

$$\underline{d}^i_j = \begin{bmatrix} L_{p_j} h_i \\ L_{p_j} L_f h_i \\ \vdots \\ -\displaystyle\sum_{k=0}^{\sigma_i-2} c^i_{k+1} L_{p_j} L^k_f h_i \end{bmatrix}, \quad 1 \le j \le n,$$

$$C^i = [1, 0, \dots, 0]_{1 \times \sigma_i}.$$

By simple matrix calculations, it can be verified that

$$C^i (\underline{A}^i)^{-1} \underline{D}^i(x) = 0, \tag{8.49}$$

for all $x \in R^n$. To this end, it follows from (8.48) and (8.49) that

$$y_i = C^i (\underline{A}^i)^{-1} \left[\dot{\xi}^i - \underline{D}^i(x) w \right] = C^i (\underline{A}^i)^{-1} \dot{\xi}^i. \tag{8.50}$$

It can be concluded from (8.50) that the disturbances are compensated from the output channel of system (8.34) with arbitrary DRDs in a static way. Particularly, if the system has a steady state, it can be obtained that the steady state value of output y_i satisfies $y_{is} = 0$.

A novel NDOBC law is constructed by replacing disturbance vector d in the disturbance decoupling control law (8.39) with disturbance estimation vector $\hat{d}$ of NDO (8.36).

Theorem 8.4 *Consider a MIMO nonlinear system (8.34) with disturbances which have arbitrary DRDs satisfying Assumptions 8.3–8.4, and also suppose that observer gain $l(x)$ of NDO (8.36) is selected such that system (8.38) is asymptotically stable. A generalized NDOBC law which compensates the disturbances in the output channels of system (8.34) is given by*

$$u = A^{-1}(x)[-b(x) + v + \Gamma(x)\hat{d}], \tag{8.51}$$

where $\hat{d}$ is the disturbance estimation vector by NDO (8.36), $A(x)$, $b(x)$, v, and $\Gamma(x)$ are the same as those defined in (8.35) and (8.39).

Proof Under the new group of coordinate transformation $\Phi(x)$ in (8.41), nonlinear system (8.34) has been reformulated as those in (8.43)–(8.45). Substituting control law (8.51) into (8.45), gives

$$\dot{\xi} = v + \Gamma(x)\hat{d} + D(x)d, \tag{8.52}$$

or equivalently expressed as

$$\dot{\xi}^i_{\sigma_i} = v_i + \sum_{k=1}^{n}(\gamma_{ik} + L_{p_k} L_f^{\sigma_i-1} h_i)d_k + \sum_{k=1}^{n}\gamma_{ik}(\hat{d}_k - d_k), \tag{8.53}$$

for $1 \leq i \leq m$. Combining (8.53) with (8.43), subsystem Π_i is rewritten as

$$\Pi_i : \begin{cases} \dot{\xi}^i = \underline{A}^i\xi^i + \underline{D}^i(x)d - \underline{\Gamma}^i(x)e, \\ y_i = C^i\xi^i, \end{cases} \tag{8.54}$$

for $1 \leq i \leq m$, where $\underline{A}^i$, $\underline{D}^i(x)$, C^i have been given in (8.48), and

$$\underline{\Gamma}^i(x) = \begin{bmatrix} 0 & 0 & \cdots & 0 \\ \vdots & \vdots & \ddots & \vdots \\ 0 & 0 & \cdots & 0 \\ \gamma_{i1}(x) & \gamma_{i2}(x) & \cdots & \gamma_{in}(x) \end{bmatrix}.$$

It follows from (8.54) and (8.49) that

$$\begin{aligned} y_i &= C^i(\underline{A}^i)^{-1}\left[\dot{\xi}^i - \underline{D}^i(x)w + \underline{\Gamma}^i(x)e\right] \\ &= C^i(\underline{A}^i)^{-1}\left[\dot{\xi}^i + \underline{\Gamma}^i(x)e\right]. \end{aligned} \tag{8.55}$$

(8.55) implies that the disturbances are decoupled from the output channel of non-linear system (8.34) in an asymptotical way. It is derived from (8.55) that

$$\begin{aligned} |y_i| &\leq \| C^i(\underline{A}^i)^{-1} \| \cdot \left(\| \dot{\xi}^i \| + \| \underline{\Gamma}^i(x) \| \cdot \| e \|\right) \\ &\leq \| C^i(\underline{A}^i)^{-1} \| \cdot \left(\| \dot{\xi}^i \| + L_1 \| e \|\right), \end{aligned} \tag{8.56}$$

where L_1 is a constant such that $\| \underline{\Gamma}^i(x) \| \leq L_1$ over the bounded area of x. Since Lemma 8.5 shows that $\lim_{t\to\infty} e(t) = 0$, the following condition, which represents the steady state of output, can be obtained by taking limits on both sides of (8.56),

$$\begin{aligned} \lim_{t\to\infty} |y_i(t)| &\leq \| C^i(\underline{A}^i)^{-1} \| \\ &\quad \times \left(\lim_{t\to\infty} \| \dot{\xi}^i(t) \| + L_1 \lim_{t\to\infty} \| e(t) \|\right) \\ &= 0. \end{aligned} \tag{8.57}$$

This completes the proof.

Remark 8.7 *If the disturbances in (8.34) are matched ones, the result in Theorem 8.4 reduces to the existing DOBC. Particularly, $g(x) = p(x)$ implies that $\Gamma(x) = -A(x)$,* and the resultant control law (8.51) becomes $u = A^{-1}(x)[-b(x) + v] - \hat{d}$, which is the same as that of the existing method [10].

Remark 8.8 *In the absence of disturbances, the disturbance estimate by (8.36) satisfies $\hat{d}(t) \equiv 0$ if the initial state of NDO (8.36) is set as $z(0) = -\lambda(x(0))$. In this case, the control performance under the proposed control law (8.51) recovers to that under the baseline feedback control law, which implies that the property of nominal performance recovery is obtained by the proposed NDOBC.*

8.4.2 Stability Analysis

Assumption 8.5 *The zero dynamics of nonlinear system (8.34) in the absence of disturbances, i.e., $\dot{x} = f(x) + g(x)u$, under the baseline feedback control law $u = A^{-1}(x)[-b(x) + v]$ are locally asymptotically stable at x_0.*

This is a necessary condition for stability analysis of the closed-loop MIMO nonlinear system under the baseline controller [106].

Lemma 8.6 [106] *Consider a system*

$$\begin{cases} \dot{\xi} = \bar{A}\xi + p(\xi, \eta), \\ \dot{\eta} = q(\xi, \eta), \end{cases} \tag{8.58}$$

and suppose that $p(0, \eta) = 0$ for all η near 0 and $\frac{\partial p}{\partial \xi}(0, 0) = 0$. If $\dot{\eta} = f(0, \eta)$ has an asymptotically stable equilibrium at $\eta = 0$ and the eigenvalues of $\bar{A}$ all have negative real parts, then the system (8.58) has an asymptotically stable equilibrium at $(\xi, \eta) = (0, 0)$.

The stability of the resultant closed-loop system is established by the following theorem.

Theorem 8.5 *Nonlinear system (8.34) under the proposed NDOBC law (8.51) is locally ISS around x_0 if the following conditions are satisfied*

  (i) Assumptions 8.2, 8.4 and 8.5 are satisfied.
  (ii) Control parameters c_k^i in control law (8.51) are selected such that polynomials $p_0^i(s)$ in (8.40) are Hurwitz stable.
 (iii) Observer gain $l(x)$ is selected such that system (8.38) is asymptotically stable.
 (iv) The disturbance compensation gain is selected such that $g(x)A^{-1}(x)\Gamma(x) + p(x)$ is continuously differentiable around x_0.

Proof Taking into account observer dynamics (8.37) and substituting NDOBC law (8.51) into system (8.34), an augmented closed-loop system is obtained, described by

$$\begin{cases} \dot{x} = F(x, e, d) \\ \dot{e} = G(e, \dot{d}) \end{cases} \tag{8.59}$$

where

$$F(x, e, d) = f(x) + g(x) A^{-1}(x)[-b(x) + v - \Gamma(x)e]$$
$$+ [g(x) A^{-1}(x)\Gamma(x) + p(x)]d, \tag{8.60}$$

$$G(e, \dot{d}) = -l(x) p(x)e + \dot{d}. \tag{8.61}$$

Under the new coordinate transformations $(\xi^T, \eta^T)^T$, the closed-loop system consisting of system $\dot{x} = f(x) + g(x)u$ and baseline feedback control law $u = A^{-1}(x)[-b(x) + v]$ is given by

$$\begin{cases} \dot{\xi} = \overline{A}\xi, \\ \dot{\eta} = q(\xi, \eta), \end{cases} \tag{8.62}$$

where $\overline{A} = diag\{\underline{A}^1, \ldots, \underline{A}^m\}$. It is easy to verify that closed-loop system (8.62) has the form of system (8.58). Assumption 8.5 and the condition (ii) imply that conditions of Lemma 8.6 are satisfied for system (8.62), and it can be concluded from the lemma that system (8.62) (or equivalently the system $\dot{x} = F(x, 0, 0)$) has an asymptotically stable equilibrium at $(\xi, \eta) = (0, 0)$ (or $x = 0$).

Let $X = [x^T, e^T]^T$, augmented closed-loop system (8.59) is rewritten as

$$\dot{X} = \underline{F}(X) + \underline{G}(X)\underline{d}, \tag{8.63}$$

where

$$\underline{F}(X) = \begin{bmatrix} F(x, e, 0) \\ G(e, 0) \end{bmatrix},$$

$$\underline{G}(X) = \begin{bmatrix} g(x) A^{-1}(x)\Gamma(x) + p(x) & 0 \\ 0 & I_{n\times n} \end{bmatrix},$$

$$\underline{d} = \begin{bmatrix} d \\ \dot{d} \end{bmatrix}.$$

According to the asymptotical stability theorem of cascade-connected systems (see Corollary 10.3.2 in [119]), system $\dot{X} = \underline{F}(X)$ is locally asymptotically stable at $X = 0$.

Combining the above result with condition (iv), it follows from Lemma 5.4 in [113] that augmented closed-loop system (8.63) is locally ISS, which completes the proof of this theorem.

8.5 An Illustrative Example

Consider a MIMO nonlinear system depicted by

$$
\begin{cases}
\dot{x}_1 = -x_1 + x_1 x_2 + x_3 + w_1, \\
\dot{x}_2 = \sin x_1 + x_2^2 + x_4 + w_2, \\
\dot{x}_3 = x_4 + u_1, \\
\dot{x}_4 = u_2, \\
\dot{y}_1 = x_1, \ y_2 = x_2.
\end{cases}
\tag{8.64}
$$

The IRDs and DRDs of system (8.64) are calculated as $(\sigma_1, \sigma_2) = (2, 2)$, and $(\nu_1, \nu_2) = (1, 1)$, respectively. This is the case where the DRDs of the system are strictly lower than the IRDs, and the conditions in Remarks 8.1 and 8.2 are not satisfied.

The NDOBC law for system (8.64) is constructed by (8.51) with some derivative calculations. The control parameters are selected as

$$
c_0^1 = 20, \ c_1^1 = 9, \ c_0^2 = 24, \ c_1^2 = 10.
$$

The gain matrix in the NDO (8.36) is selected as

$$
\lambda(x) = [50x_1, 50x_2]^T.
$$

To evaluate the effectiveness of the proposed method, both the baseline control and integral control methods are employed in the simulation studies for the purpose of comparison. The baseline controller is given as

$$
u = A^{-1}(x)[-b(x) + v].
\tag{8.65}
$$

The integral control is designed as

$$
u = A^{-1}(x)[-b(x) + v + u_{int}],
\tag{8.66}
$$

where

$$
u_{int} = \left[-c_i^1 \int (y_1 - r_1) dt, \ -c_i^2 \int (y_2 - r_2) dt \right]^T,
$$

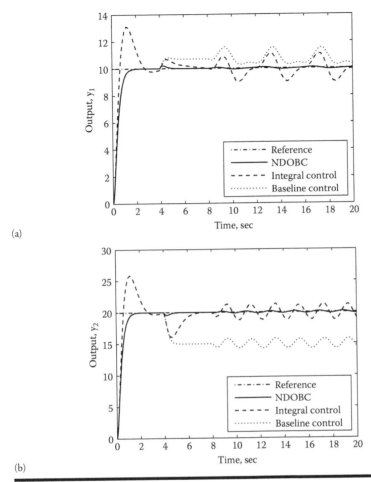

(a)

(b)

Figure 8.1 **Input profiles of system (8.64) in the presence of unknown external disturbances under control laws (8.51) (solid line), (8.65) (dotted line), and (8.66) (dashed line).**

with r_1 and r_2 the reference signals, c_i^1 and c_i^2 are the integral coefficients, which are chosen as $c_i^1 = 25$ and $c_i^2 = 30$ for simulation studies.

The reference signals are selected as $r_1 = 10$ and $r_2 = 20$, respectively. The unknown external disturbances,

$$\begin{cases} w_1(t) = 0, \ w_2(t) = 0, & \text{for } 0 \leq t < 4, \\ w_1(t) = 2, \ w_2(t) = -3, & \text{for } 4 \leq t < 8, \\ w_1(t) = 0.6\sin(\tfrac{\pi}{2}t) + 2, \ w_2(t) = -0.8\sin(\pi t + \tfrac{\pi}{2}) - 3, & \text{for } 8 \leq t \leq 20, \end{cases} \tag{8.67}$$

are considered to be imposed on system (8.64).

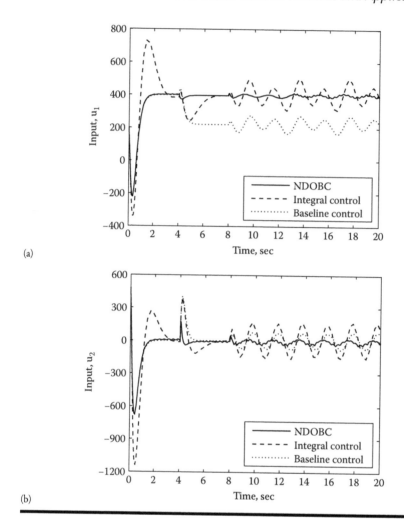

(a)

(b)

Figure 8.2 **Output response curves of system (8.64) in the presence of unknown external disturbances under control laws (8.51) (solid line), (8.65) (dotted line), and (8.66) (dashed line). The reference signals are denoted by dash-dotted line.**

The output response curves of system (8.64) subject to disturbances (8.67) under the three controllers are shown in Figure 8.2. The corresponding input profiles are shown by Figure 8.1. It is observed from Figure 8.2 that the baseline controller achieves excellent tracking performance, but quite poor disturbance attenuation performance. The integral control method improves the disturbance rejection performance to some extent, however, it brings certain undesired control performance, such as unsatisfactory overshoots. This implies that the disturbance rejection performance of integral control is achieved at the price of sacrificing its nominal control performance.

A brief observation of Figure 8.2 shows that the proposed NDOBC method exhibits a promising disturbance attenuation and reference tracking performance. It is also observed from Figures 8.2 and 8.1 that the response curves under the NDOBC are overlapped with those under the baseline control method during the first 4 seconds when there are no disturbances imposed on the system, which is a clear evidence that *the property of nominal performance recovery is retained for the NDOBC method.*

8.6 Summary

The disturbance attenuation problem of a nonlinear system with arbitrary DRD via a DOBC approach has been addressed in this chapter by designing a new disturbance compensation gain matrix in the control law. It has been proved that the disturbances in this case could be eliminated from the output channels by the newly proposed NDOBC with properly chosen control parameters. The proposed NDOBC method has retained the major feature of the DOBC method, that is, the ability of recovering its nominal control performance, which has been verified by simulation studies of both numerical and application examples.

Chapter 9

Linear/Nonlinear Disturbance Observer-Based Sliding Mode Control for Systems with Mismatched Uncertainties

9.1 Introduction

Sliding-mode control has been widely studied for over 50 years and widely employed in industrial applications due to its conceptual simplicity, and in particular, powerful ability to reject disturbances and plant uncertainties [120, 121, 122]. It is noticed that most of the existing results on sliding surface design are concentrated on the matched uncertainties attenuation since the sliding motion of traditional SMC is only insensitive to matched uncertainties [123] but sensitive to mismatched uncertainties.

Due to the significance of attenuating mismatched uncertainties in practical control applications, many authors work on the sliding surface design for uncertain systems subject to mismatched disturbances (see, e.g., Refs. [123, 124, 125, 126, 127, 128, 129] and references therein). In general, these SMC methods can be classified into the following two categories.

The first category mainly focuses on the stability (or robust stability) of various systems under mismatched uncertainties using some classical control design tools, such as Riccati approach [124, 125] and LMI-based approach [123, 126, 127]. The mismatched uncertainties considered by those methods must be H_2 norm-bounded, that is, the mismatched uncertainties must belong to vanishing uncertainties, which is generally not a reasonable assumption for practical systems since many engineering systems may suffer from mismatched disturbances that do not necessarily satisfy the condition of with a bounded H_2 norm. Taking the permanent magnet synchronous motor as an example, the lumped uncertainties therein may have nonzero steady-state values and thus do not have bounded H_2 norms [101].

The second category is referred to as integral sliding-model control (I-SMC) [125, 129]. The idea behind I-SMC is that a high-frequency switching gain is designed to force the states to achieve integral sliding surface, and then the integral action in the sliding surface drives the states to the desired equilibrium in the presence of mismatched uncertainties. Compared with the first category SMC for mismatched uncertainties, the I-SMC method is more practical due to its simplicity and robustness and has been reported to be applied to many systems [125, 129]. However, it is well known that integral action always brings some adverse effects to the control systems, such as large overshoot and long settling time.

Note that the above two categories of SMC methods deal with the mismatched uncertainties in a robust way, which implies that the uncertainty attenuation ability is achieved at the price of sacrificing its nominal control performance. Moreover, the chattering problem in these methods is still a severe problem to be solved. In this chapter, novel sliding-mode control methods are addressed to counteract mismatched uncertainties in the system via a nonlinear disturbance observer. The mismatched uncertainties under consideration are possibly nonvanishing and do not necessarily satisfy the condition of H_2 norm-bounded. By designing a new sliding surface based on the disturbance estimation, the system states can be driven to the desired equilibrium asymptotically by sliding motion along the sliding surface even under mismatched disturbance. A discontinuous control law with a high frequency switching gain is then designed to force the initial states to reach the designed sliding surface. There are mainly two remarkable features of the proposed method. First, the high frequency switching gain in the proposed control law is only required to be designed larger than the bound of the disturbance estimation error rather than that of the disturbance, which substantially alleviates the chattering problem. Second, the proposed method retains its nominal performance since the disturbance observer serves like a patch to the baseline controller and does not cause any adverse effects on the system in the absence of disturbances and uncertainties. Part contents of this chapter refer to [18].

9.2 Linear Disturbance Observer-Based Sliding-Mode Control

9.2.1 Problems of the Existing SMC Methods

Consider the following second-order system subject to mismatched disturbances

$$\begin{cases} \dot{x}_1 = x_2 + d(t), \\ \dot{x}_2 = a(x) + b(x)u, \\ y = x_1, \end{cases} \tag{9.1}$$

where x_1 and x_2 the states, u the control input, $d(t)$ the disturbances, and y the output.

Assumption 9.1   *The disturbances in system (9.1) are bounded by $d^* = \sup\limits_{t>0} |d(t)|$.*

The sliding-mode surface and control law of the traditional SMC are generally designed as follows

$$s = x_2 + cx_1, \; u = -b^{-1}(x)[a(x) + cx_2 + ksgn(s)]. \tag{9.2}$$

Combining (9.1)–(9.2) yields

$$\dot{s} = -ksgn(s) + cd(t). \tag{9.3}$$

Equation (9.3) implies that the states of system (9.1) initially outside the sliding surface will reach the sliding surface $s = 0$ in a finite time as long as the switching gain in control law (9.2) is chosen such that $k > cd^*$. Taking into account the condition $s = 0$ in (9.2), the sliding motion is obtained by

$$\dot{x}_1 = -cx_1 + d(t). \tag{9.4}$$

Remark 9.1   *Equation (9.4) shows that the states can not reach the desired equilibrium point even control law (9.2) can force the system states to reach the sliding surface in finite time. This is the essential reason why the traditional SMC is only insensitive to matched disturbance but sensitive to mismatched one.*

An efficient solution for attenuating mismatched disturbances is known as integral sliding-model control (I-SMC), which defines the following sliding-mode surface

$$s = x_2 + c_1 x_1 + c_2 \int x_1. \tag{9.5}$$

An integral SMC law is generally designed as

$$u = -b^{-1}(x)[a(x) + c_1 x_2 + c_2 x_1 + k \, sgn(s)]. \tag{9.6}$$

Combining (9.1), (9.5), and (9.6) yields

$$\dot{s} = -k \, sgn(s) + c_1 d(t). \tag{9.7}$$

The states of system (9.1) initially outside sliding surface $s = 0$ will reach surface $s = 0$ in (9.5) in finite time as long as the switching gain in control law (9.6) is chosen such that $k > c_1 d^*$. Taking into account the condition $s = 0$ gives

$$\ddot{x}_1 + c_1 \dot{x}_1 + c_2 x_1 = d(t), \tag{9.8}$$

which implies that the state can reach the desired equilibrium point asymptotically if the system has reached sliding surface $s = 0$ in finite time, and the disturbance has a constant steady state value, that is, $\lim_{t \to \infty} d(t) = 0$.

Remark 9.2 *The I-SMC method is effective to remove the offset caused by mismatched disturbance but in a robust way. To this end, the integral action lies in the I-SMC method always brings several adverse effects to the control performance, such as introducing overshoot and destroying its nominal control performance. This will be shown in the latter simulation example.*

9.2.2 Novel SMC Method Based on a Disturbance Observer

9.2.2.1 Control Design

Let $x = [x_1, x_2]^T$, then system (9.1) is expressed as

$$\begin{cases} \dot{x} = f(x) + g_1(x)u + g_2 d, \\ y = x_1, \end{cases} \tag{9.9}$$

where $f(x) = [x_2, a(x)]^T$, $g_1(x) = [0, b(x)]^T$, and $g_2 = [1, 0]^T$.

A nonlinear disturbance observer (NDO), which can estimate the disturbance in (9.9), is introduced and depicted by [40, 2]

$$\begin{cases} \dot{p} = -lg_2 p - l[g_2 lx + f(x) + g_1(x)u], \\ \hat{d} = p + lx, \end{cases} \tag{9.10}$$

where $\hat{d}$, p, and l are the estimation of disturbances, the internal state of the nonlinear observer, the observer gain to be designed, respectively.

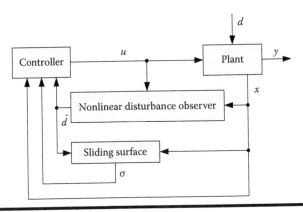

Figure 9.1 **Control configuration of the proposed disturbance observer based sliding-mode control method.**

A novel sliding-mode surface for system (9.1) under mismatched disturbance is defined by introducing the disturbance estimation of (9.10)

$$s = x_2 + cx_1 + \hat{d}. \tag{9.11}$$

where $\hat{d}$ is the disturbance estimation given by NDO (9.10), and $c > 0$ is a control parameter to be designed.

The proposed disturbance observer based sliding-mode control law is designed as

$$u = -b^{-1}(x)[a(x) + c(x_2 + \hat{d}) + k\,sgn(s)], \tag{9.12}$$

where k is the switching gain to be designed.

The control configuration of the proposed disturbance observer based sliding-mode control is given by Figure 9.1.

9.2.2.2 Stability Analysis

Assumption 9.2 *The derivative of the disturbances in system (9.1) is bounded and satisfies* $\lim_{t \to \infty} \dot{d}(t) = 0$.

Lemma 9.1 [40] *Suppose that Assumptions 9.1 and 9.2 are satisfied for system (9.1). Disturbance estimation $\hat{d}$ of NDO (9.10) can track disturbances d of system (9.1) asymptotically if observer gain l is chosen such that $lg_2 > 0$ holds, which implies that*

$$\dot{e}_d(t) + lg_2 e_d(t) = 0, \tag{9.13}$$

is globally asymptotically stable, where $e_d(t) = d(t) - \hat{d}(t)$ is the disturbance estimation error.

Assumption 9.3 *The disturbance estimation error in (9.13) is bounded by $e_d^* = \sup_{t>0} |e_d(t)|$.*

Lemma 9.2 *[113] Consider a nonlinear system $\dot{x} = F(x, w)$ which is input-to-state stable (ISS). If the input satisfies $\lim_{t\to\infty} w(t) = 0$, then the state $\lim_{t\to\infty} x(t) = 0$.*

Theorem 9.1 *Suppose that Assumptions 9.1-9.3 are satisfied for system (9.1). Considering system (9.1) under proposed control law (9.12), the closed-loop system is asymptotically stable if the switching gain in control law (9.12) is designed such that $k > (c + lg_2)e_d^*$ and observer gain l is chosen such that $lg_2 > 0$ holds.*

Proof Taking the derivative of sliding surface s (9.11) along system (9.1) gives

$$\dot{s} = a(x) + b(x)u + c[x_2 + d(t)] + \dot{d}. \tag{9.14}$$

Substituting control law (9.12) into (9.14), one obtains

$$\dot{s} = -ksgn(s) + c[d(t) - \hat{d}(t)] + \dot{d}. \tag{9.15}$$

It can be derived from (9.10) that

$$\dot{\hat{d}}(t) = -lg_2[\hat{d}(t) - d(t)]. \tag{9.16}$$

Substituting (9.16) into (9.15) yields

$$\dot{s} = -ksgn(s) + (c + lg_2)e_d(t). \tag{9.17}$$

Consider a candidate Lyapunov function as

$$V(s) = s^2/2. \tag{9.18}$$

Taking derivative of $V(s)$ (9.18) along (9.17) gives

$$\begin{aligned} \dot{V} &= -k|s| + (c + lg_2)e_d(t)s \\ &\leq -[k - (c + lg_2)e_d^*]|s| \\ &= -\sqrt{2}[k - (c + lg_2)e_d^*]V^{\frac{1}{2}}. \end{aligned} \tag{9.19}$$

Taking into account the given condition $k > (c + lg_2)e_d^*$, it can be derived from (9.19) that system states will reach the defined sliding surface $s = 0$ (9.11) in finite time. The condition $s = 0$ implies that

$$\dot{x}_1 = -cx_1 + d(t) - \hat{d}(t). \tag{9.20}$$

Combining (9.20) with the observer dynamics, one obtains

$$\begin{cases} \dot{x}_1 = -cx_1 + e_d, \\ \dot{e}_d = -lg_2 e_d + \dot{d}, \\ x_2 = -cx_1 - \hat{d}. \end{cases} \tag{9.21}$$

Under the given conditions that $c > 0$ and $lg_2 > 0$, it can be shown that the following system

$$\begin{cases} \dot{x}_1 = -cx_1 + e_d, \\ \dot{e}_d = -lg_2 e_d \end{cases} \tag{9.22}$$

is exponentially stable. With this result, it is concluded from Lemma 5.5 in [113] that system

$$\begin{cases} \dot{x}_1 = -cx_1 + e_d, \\ \dot{e}_d = -lg_2 e_d + \dot{d} \end{cases} \tag{9.23}$$

is ISS. Based on the condition in Assumption 9.2, it is concluded from Lemma 9.2 that the states of system (9.23) satisfy $\lim_{t\to\infty} x_1(t) = 0$ and $\lim_{t\to\infty} e_d(t) = 0$. This implies that the system states will converge to the desired equilibrium point asymptotically with the proposed control law.

Remark 9.3 *To guarantee stability, the switching gains of traditional SMC, I-SMC, and DOB-SMC have to be designed such that $k > c|d|$, $k > c_1|d|$, and $k > (c + lg_2)|d - \hat{d}|$, respectively. Since the disturbance has been precisely estimated by disturbance observer, the magnitude of estimation error $|d - \hat{d}|$, which is expected to converge to 0, can be kept much smaller than the magnitude of disturbance d. The readers can also refer to [109] for the same argument. Thus the switching gain of the proposed method can be designed much smaller than those of traditional SMC and I-SMC methods, and the chattering problem can be alleviated to some extent, which will be shown by the later simulation studies.*

Remark 9.4 *In the absence of disturbance, it is derived from (9.16) that*

$$\dot{\hat{d}}(t) = -lg_2 \hat{d}(t), \tag{9.24}$$

which implies that $\hat{d}(t) \equiv 0$ if its initial value is selected as $\hat{d}(0) = 0$. Sliding surface (9.11) and control law (9.12) of the proposed method reduces to those of the traditional SMC in (9.2). This implies that the nominal control performance of the proposed method is retained. Such excellent feature will be shown by the numerical example in next section.

Table 9.1 Control Parameters for the Numerical Example in Case 1

Methods	Control Parameters
SMC	$c = 5, k = 3$
I-SMC	$c_1 = 5, c_2 = 6, k = 3$
DOB-SMC	$c = 5, k = 3, \lambda(x) = 6x_1$

9.2.3 An Illustrative Example

Considering the following system for simulation studies

$$\begin{cases} \dot{x}_1 = x_2 + d(t), \\ \dot{x}_2 = -2x_1 - x_2 + e^{x_1} + u, \\ y = x_1. \end{cases} \tag{9.25}$$

9.2.3.1 Nominal Performance Recovery

Consider the initial states of system (9.25) as $x(0) = [-1, 1]^T$. A step external disturbance $d = 0.5$ is imposed on the system at $t = 6$ sec. Both the traditional SMC and integral SMC methods are employed in the control design for system (9.25) for simulation studies. The control parameters of all the three control methods are listed in Table 9.1.

It can be observed from Figures 9.2 and 9.3 that the states and control input responses under the proposed method are the overlapped with those under the baseline SMC method during the first 6 sec, which is an evidence as stated in Remark 9.4 that the nominal control performance of the proposed method is retained. It is also noticed that integral action of the I-SMC method causes adverse control effects such as undesired overshoot and unsatisfactory settling time.

It can be observed from Figure 9.2 that the traditional SMC method fails to drive the states to the desired equilibrium in the presence of mismatched disturbance, which indicates the results in Remark 9.1 that traditional SMC is sensitive to mismatched disturbance. Both the proposed method and the I-SMC method can finally suppress the mismatched disturbance, but the proposed method exhibits a much quicker convergence rate than the I-SMC method.

9.2.3.2 Chattering Reduction

As clearly shown in Figure 9.3, all the three controllers have resulted in substantial chattering since a large switching gain has been designed to attenuate the disturbance.

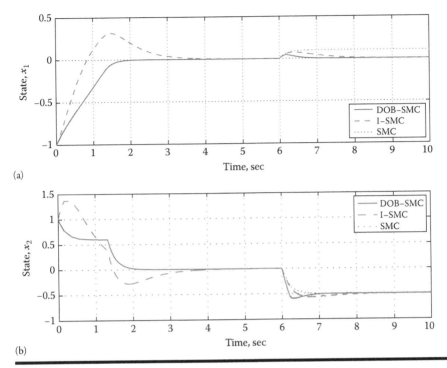

(a)

(b)

Figure 9.2 **State variables in simulation scenario 1 (nominal performance recovery).**

Actually, as stated in Remark 9.3, the proposed method can alleviate the chattering by setting a relatively smaller switching gain. In this part, we try to show such feature by simulation studies.

Consider the initial states of system (9.25) as $x(0) = [0, 0]^T$ now. A step external disturbance $d = -0.5$ is imposed on the system at $t = 2$ sec. The control parameters of all the three control methods are listed in Table 9.2.

It can be observed from Figures 9.4–9.5 that the chattering is reduced for the proposed DOB-SMC method by decreasing the switching gain. In this simulating scenario, the switching gain k is decreased from 3 to 1 for all the three methods. However, as clearly shown in the simulation results, the proposed method still obtains fine control performance with a much smaller chattering than that in simulation scenario 1. When we try to reduce the chattering under the SMC and I-SMC methods by decreasing the switching gain, it is observed that such two methods fail to reject the disturbances of system effectively. The results in this simulation scenario indicate the conclusions given in Remark 8.3.

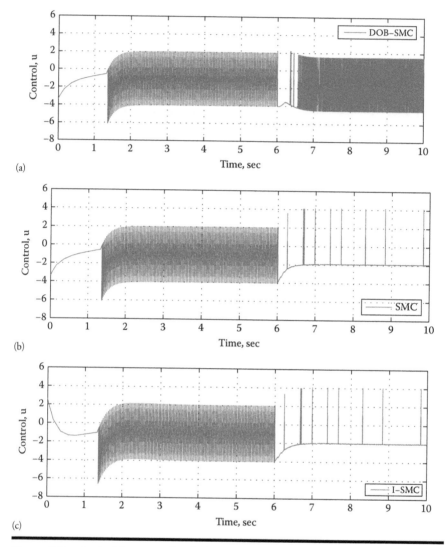

(a)

(b)

(c)

Figure 9.3 Control input in simulation scenario 1 (nominal performance recovery).

Table 9.2 Control Parameters for the Numerical Example in Case 2

Methods	Control Parameters
SMC	$c = 5, k = 1$
I-SMC	$c_1 = 5, c_2 = 6, k = 1$
DOB-SMC	$c = 5, k = 1, \lambda(x) = 6x_1$

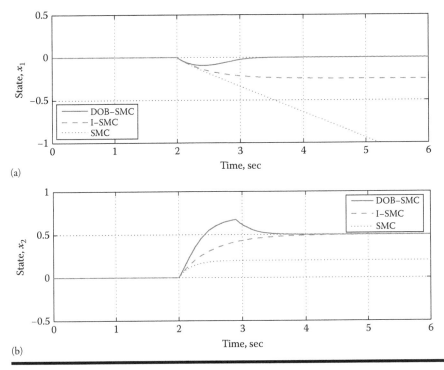

(a)

(b)

Figure 9.4 State variables in simulation scenario 2 (Chattering reduction).

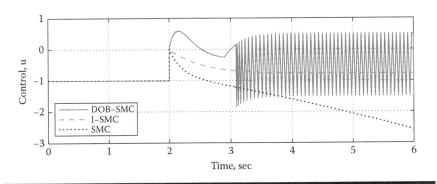

Figure 9.5 Control input in simulation scenario 2 (Chattering reduction).

9.3 Nonlinear Disturbance Observer-Based Nonsingular Terminal Sliding-Mode Control

9.3.1 Problem of the Existing NTSMC Methods

Consider the following second-order system under mismatched disturbance, depicted by

$$
\begin{cases}
\dot{x}_1 = x_2 + d(t), \\
\dot{x}_2 = a(x) + b(x)u + d_0(t), \\
y = x_1,
\end{cases}
\tag{9.26}
$$

where $x = [x_1, x_2]^T$ is the state vector, u is the control input, y is the controlled output, $d_0(t)$ and $d(t)$ denote the matched and mismatched disturbances, respectively. $a(x)$ and $b(x) \neq 0$ are smooth nonlinear functions in terms of x.

For the existing NTSMC method, the nonlinear sliding mode surface is usually defined as follows

$$
s = x_1 + \frac{1}{\beta} x_2^{p/q},
\tag{9.27}
$$

where $\beta > 0$ is a design constant, p and q are positive odd integers which satisfy the condition that $1 < p/q < 2$.

The NTSMC law is usually designed as follows

$$
u = -b^{-1}(x) \left[a(x) + \beta \frac{q}{p} x_2^{2-p/q} + k \mathrm{sgn}(s) \right],
\tag{9.28}
$$

where k is the switching gain to be designed.

Combining (9.27) and (9.28), gives

$$
\dot{s} = -\frac{1}{\beta} \frac{p}{q} x_2^{p/q-1} \left[k \mathrm{sgn}(s) - d_0(t) \right] + d(t).
\tag{9.29}
$$

In the absence of mismatched disturbances, i.e., $d(t) = 0$, it can be concluded from [130] that conventional NTSMC law (9.28) with appropriately chosen parameters (actually $k > \sup |d_0(t)|$) can drive arbitrary initial states of system (9.26) to the equilibrium point $x = 0$ in finite time, which implies that the conventional NTSMC method is insensitive to matched disturbances.

However, in the presence of mismatched disturbances, i.e., $d(t) \neq 0$, two problems appear for the conventional NTSMC method. Firstly, it is not easy to determine the switching gain k such that the states of system (9.26) initially outside the sliding surface will reach the sliding surface $s = 0$ in finite time. Secondly, even if the sliding surface $s = 0$ is reached, the system dynamics are determined by the following nonlinear differential equation

$$\dot{x}_1 = -x_2 + d(t) = -\beta^{q/p} x_1^{q/p} + d(t), \tag{9.30}$$

which implies that the output $y = x_1$ of system (9.26) is affected by the mismatched disturbances $d(t)$, and does not converge to zero in finite time. To this end, it is imperative to address the disturbance rejection problem of the NTSMC design method in the case of mismatching condition.

9.3.2 Novel NTSMC Method Based on a Finite-Time Disturbance Observer

9.3.2.1 Finite-Time Disturbance Observer

For the sake of simplicity, we consider the mismatched disturbance attenuation problem for system (9.26) firstly. The matched disturbance case will be discussed later in Remark 9.6. To present the main result of the paper, a finite-time disturbance observer is firstly addressed to estimate the disturbance in system (9.26).

Assumption 9.4 *The mismatched disturbance in system (9.26) is rth differentiable, and $d^{[r]}$ has a Lipshitz constant L.*

A finite-time disturbance observer (FTDO) for system (9.26) is constructed as [131, 132]

$$
\begin{cases}
\dot{z}_0 = v_0 + x_2, \ \dot{z}_1 = v_1, \ \ldots, \ \dot{z}_{r-1} = v_{r-1}, \ \dot{z}_r = v_r, \\
v_0 = -\lambda_0 L^{1/(r+1)} |z_0 - x_1|^{r/(r+1)} \mathrm{sgn}(z_0 - x_1) + z_1, \\
v_1 = -\lambda_1 L^{1/r} |z_1 - v_0|^{(r-1)/r} \mathrm{sgn}(z_1 - v_0) + z_2, \\
\ \vdots \\
v_{r-1} = -\lambda_{r-1} L^{1/2} |z_{r-1} - v_{r-2}|^{1/2} \mathrm{sgn}(z_{r-1} - v_{r-2}) + z_r, \\
v_r = -\lambda_r L \mathrm{sgn}(z_r - v_{r-1}), \\
\hat{x}_1 = z_0, \ \hat{d} = z_1, \ \ldots, \ \hat{d}^{[r-1]} = z_r,
\end{cases}
\tag{9.31}
$$

where $\lambda_i > 0 (i = 0, 1, \ldots, r)$ is the observer coefficients to be designed, $\hat{x}, \hat{d}, \ldots,$ $\hat{d}^{[r-1]}$ are the estimates of $x, d, \ldots, d^{[r-1]}$, respectively.

Combining (9.26) with (9.31), the observer estimation error is governed by

$$
\begin{cases}
\dot{e}_0 = -\lambda_0 L^{1/(r+1)}|e_0|^{r/(r+1)}\text{sgn}(e_0) + e_1, \\
\dot{e}_1 = -\lambda_1 L^{1/r}|e_1 - \dot{e}_0|^{(r-1)/r}\text{sgn}(e_1 - \dot{e}_0) + e_2, \\
\vdots \\
\dot{e}_{r-1} = -\lambda_{r-1} L^{1/2}|e_{r-1} - \dot{e}_{r-2}|^{1/2}\text{sgn}(e_{r-1} - \dot{e}_{r-2}) + e_r, \\
\dot{e}_r \in -\lambda_r L\text{sgn}(e_r - \dot{e}_{r-1}) + [-L, L],
\end{cases}
\tag{9.32}
$$

where the estimation errors are defined as $e_0 = z_0 - x_1$, $e_1 = z_1 - d$, ... , $e_{r-1} = z_{r-1} - d^{[r-2]}$, $e_r = z_r - d^{[r-1]}$.

Suppose that the mismatched disturbance in (9.26) satisfies the condition of Assumption 9.4, it follows from [131] that observer error system (9.32) is finite-time stable, that is, there is a time constant $t_f > t_0$ such that $e_i(t) = 0 (i = 0, 1, \ldots, r)$ (or equivalently $\hat{x}_1(t) = x_1(t)$, $\hat{d}(t) = d(t)$, ... , $\hat{d}^{[r-1]}(t) = d^{[r-1]}(t)$) for $t \geq t_f$.

9.3.2.2 Control Design and Stability Analysis

A novel nonlinear dynamic sliding mode surface for system (9.26) under mismatched disturbance is defined based on the disturbance estimation of (9.31), given by

$$
s = x_1 + \frac{1}{\beta}(x_2 + \hat{d})^{p/q},
\tag{9.33}
$$

where β, p, and q have been defined in (9.27), $\hat{d}$ is the disturbance estimation given by FTDO (9.31).

Theorem 9.2 *For system (9.26) with the proposed novel nonlinear sliding mode surface (9.33), if the FTDO based NTSMC law is designed as*

$$
u = -b^{-1}(x) \times \left[a(x) + \beta\frac{q}{p}(x_2 + \hat{d})^{2-p/q} + v_1 + k\text{sgn}(s)|s|^{\alpha} \right],
\tag{9.34}
$$

where $0 < \alpha < 1$, v_1 has been given in (9.31), then system output $y = x_1$ will converge to zero in finite time.

Proof For the proposed sliding surface (9.33), its derivative along system dynamics (9.26) is

$$
\begin{aligned}
\dot{s} &= \dot{x}_1 + \frac{1}{\beta}\frac{p}{q}(x_2 + \hat{d})^{p/q-1}(\dot{x}_2 + \dot{\hat{d}}) \\
&= x_2 + d + \frac{1}{\beta}\frac{p}{q}(x_2 + \hat{d})^{p/q-1}[a(x) + b(x)u + v_1] \\
&= -\frac{1}{\beta}\frac{p}{q}\bar{x}_2^{p/q-1}k\text{sgn}(s)|s|^{\alpha} - e_1,
\end{aligned}
\tag{9.35}
$$

where $\bar{x}_2 = x_2 + \hat{d}$.

Define a Lyapunov function $V(s) = \frac{1}{2}s^2$ for the sliding mode dynamics (9.35). Taking the derivative of $V(x)$ along system (9.35), yields

$$\dot{V}(s) = s\dot{s} = -\frac{1}{\beta}\frac{p}{q}\bar{x}_2^{p/q-1}k|s|^{\alpha+1} - e_1 s. \tag{9.36}$$

Since the disturbance observer error system is finite-time stable, disturbance estimation error e_1 in (9.32) is bounded by $|e_1| \leq \sigma$, where $\sigma > 0$. Consider the condition that p and q are positive odd integers, it is obtained from (9.36) that

$$\dot{V}(s) \leq \sigma|s|. \tag{9.37}$$

In the case of $|s| \geq 1$, it is followed from (9.37) that $\dot{V}(s) \leq KV(s)$, $K = 2\sigma$. In the case of $|s| < 1$, there exists a constant $L = \frac{1}{2}K$ such that $\dot{V}(s) < L$. Combining both cases together, we have

$$\dot{V}(s) \leq KV(s) + L. \tag{9.38}$$

It can be concluded from (9.38) that $V(s)$ and so s will not escape in finite time.

Since disturbance estimation e_1 in (9.32) will converge to zero in a finite time t_f, system (9.35) will reduce to

$$\dot{s} = -\rho(\bar{x}_2)k\text{sgn}(s)|s|^{\alpha}, \tag{9.39}$$

where $\rho(\bar{x}_2) = \frac{1}{\beta}\frac{p}{q}\bar{x}_2^{p/q-1}$. Next we will show that (9.39) is finite-time stable. The idea of the proof procedure is inspired from [130]. For the case of $\bar{x}_2 \neq 0$, it can be followed from $\rho(\bar{x}_2) > 0$ that (9.39) is finite-time stable. Substituting the control law (9.34) into system (9.26), yields

$$\dot{\bar{x}}_2 = -\beta\frac{q}{p}\bar{x}_2^{2-p/q} - k\text{sgn}(s)|s|^{\alpha}. \tag{9.40}$$

For $\bar{x}_2 = 0$, it is obtained that

$$\dot{\bar{x}}_2 = -k\text{sgn}(s)|s|^{\alpha}. \tag{9.41}$$

Similar to the proof in [130], it can be shown that $\bar{x}_2 = 0$ is not an attractor. Therefore, it can be concluded that system (9.39) is finite-time stable, that is, there is a time $t_r > 0$ such that $s(t) = 0$ for $t \geq t_f + t_r$.

Once the sliding surface $s = 0$ is reached, it is derived from sliding surface (9.33) and system dynamics (9.26) that

$$s = x_1 + \frac{1}{\beta}(x_2 + d)^{p/q} = x_1 + \frac{1}{\beta}\dot{x}_1^{p/q} = 0, \tag{9.42}$$

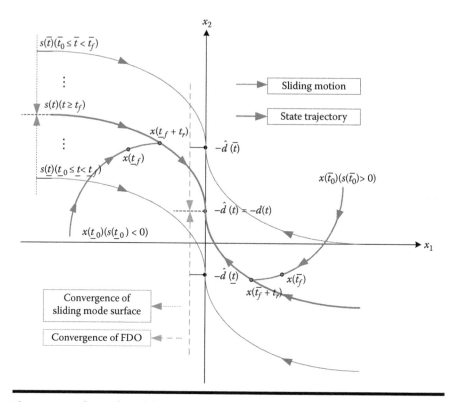

Figure 9.6   Phase plot of the system under the proposed FDO-NTSMC method.

for $t \geq t_f + t_r$. With the chosen control parameters, system (9.42) is finite-time stable, which implies that there is a time constant $t_s > 0$ such that $y(t) = x_1(t) = 0$ and $x_2(t) = -d(t)$ for $t \geq t_f + t_r + t_s$.

The phase plane plot of the system under the proposed FDO-NTSMC method is shown in Figure 9.6. The phase plane motion mainly includes the following three steps.

Step 1 ($t_0 \leq t \leq t_f$): the convergence motion of FDO dynamics occurs. The disturbance estimation will converge to its real value during this interval in finite time, that is, $\hat{d}(t) = d(t)$ for all $t \geq t_f$. The convergence motion of sliding mode surface occurs correspondingly, that is, $s(t) = x_1 + \frac{1}{\beta}(x_2 + d)^{p/q}$ for all $t \geq t_f$.

Step 2 ($t_f < t \leq t_f + t_r$): the states will reach the switching line during such interval in finite time, that is, $s(t) = 0$ for all $t \geq t_f + t_r$.

Step 3 ($t_f + t_r < t \leq t_f + t_r + t_s$): the nonsingular terminal sliding motion occurs in this step. States $(x_1, \bar{x}_2)$ in the switching line will reach zero in finite time, that is, $y(t) = x_1(t) = 0$ and $x_2(t) = -d(t)$ for all $t \geq t_f + t_r + t_s$.

Remark 9.5 (Nominal Performance Recovery.) *In the absence of disturbance, it is derived from observer error dynamics (9.32) that* $e_0(t) = e_1(t) = \ldots = e_r(t) = 0$

and $v_1(t) = 0$ if the initial values of observer states are selected as $z_0(t_0) = x_1(t_0)$ and $z_1(t_0) = \ldots = z_r(t_0) = 0$. In this case, proposed sliding surface (9.33) and control law (9.34) reduce to those of traditional NTSMC in (9.27) and (9.28). This implies that the nominal control performance of the proposed method is retained.

Remark 9.6 *Although the matched disturbance attenuation problem is not discussed above, the proposed continuous NTSMC method (9.34) can be easily extended to tackle both matched and mismatched disturbances. Actually, the extended control law can be designed as*

$$u = -b^{-1}(x) \times \left[a(x) + \hat{d}_0 + \beta \frac{q}{p}(x_2 + \hat{d})^{2-p/q} + v_1 + k\,\mathrm{sgn}(s)|s|^\alpha \right], \quad (9.43)$$

where $\hat{d}_0$ (that can be derived by another FDO [131, 132]) is the estimate of the matched disturbance d_0. The proof of this result is the same as that of Theorem 9.2, which is omitted here for space.

9.3.3 An Illustrative Example

Consider the numerical example (9.25) for simulation studies. For the purpose of comparison studies, four control approaches are adopted: the proposed method (FTO-NTSMC), the NTSMC method, the method proposed in Section 8.2 (DO-LSMC), and the integral SMC (I-SMC). The controller and observer parameters of the FDO-NTSMC method is designed as

$$\beta = 3, \ p = 5, \ q = 3, \ k = 3, \ \alpha = 1/3, \ L = 10, \ \lambda_0 = 3, \ \lambda_1 = 1.5, \ \lambda_2 = 1.1.$$

The control parameters of the NTSMC method is designed the same as those of the baseline controller of the FDO-NTSMC method. The control parameters of both the DO-SMC and the I-SMC are selected the same as those in Section 8.2.

Consider the initial states of example system as $x(0) = [1, -1]^T$. The following mismatched disturbances is supposed to impose on the system: $d(t) = 0$ for $t < 5$; $d(t) = 0.6$ for $5 \leq t < 10$; and $d(t) = 0.6 + 0.2 \sin 2\pi t$ for $10 \leq t \leq 15$. Response curves of the state variables under the four methods are shown by Figure 9.7. The corresponding control signals are shown in Figure 9.8.

It is observed from Figure 9.7 that the state responses of the proposed method are the same as those of the baseline NTSMC method during the first 5 sec, which shows the result stated in Remark 9.5 that the nominal performance recovery of the proposed method. As shown by Figure 9.7(a), although both the DO-LSMC and the I-SMC methods could asymptotically attenuate the mismatched step external disturbances, the proposed method exhibits the fastest disturbance attenuation property. In addition, the DO-LSMC and the I-SMC methods only attenuate the

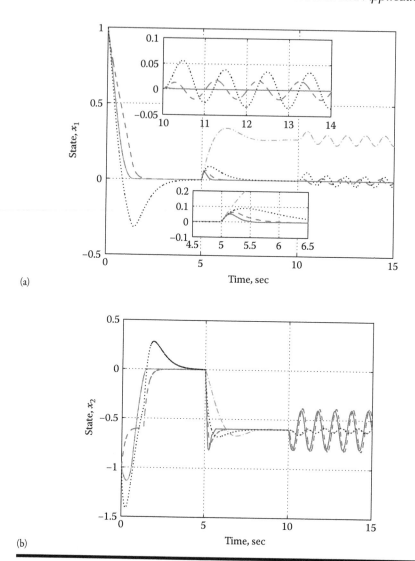

(a)

(b)

Figure 9.7 **State response curves in the presence of unknown mismatched disturbances under FDO-NTSMC (solid line), NTSMC (dash-dotted line), DO-LSMC (dashed line), and I-SMC (dotted line) methods: (a) the output/the first state,** $y = x_1$**; (b) the second state,** x_2**.**

harmonic disturbances to a specified small region, while the proposed method has removed such disturbances completely. It is observed from Figure 9.8 that the chattering problem is largely reduced by the proposed method due to the continuous control law design of (9.34).

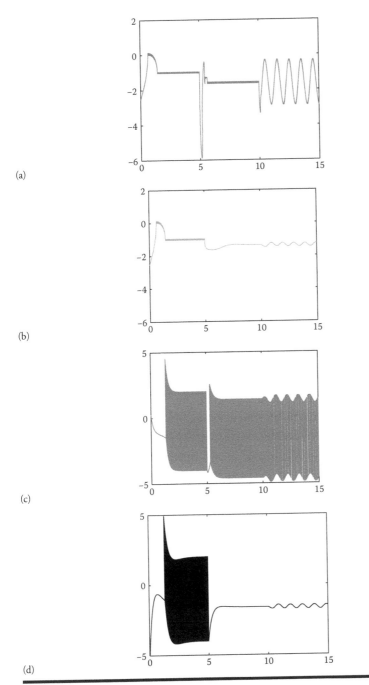

(a)

(b)

(c)

(d)

Figure 9.8 **Control signals in the presence of unknown mismatched disturbances under the four control methods: (a) FDO-NTSMC; (b) NTSMC; (c) DO-LSMC; and (d) I-SMC.**

9.4 Summary

Firstly, a novel disturbance observer based sliding-mode control approach has been proposed to attenuate the mismatched uncertainties. The main contribution here is to design a new sliding-mode surface which includes the disturbance estimation such that the sliding motion along the sliding surface can drive the states to the desired equilibrium point in the presence of mismatched uncertainties. As compared with the traditional SMC and integral SMC methods, the proposed method has exhibited two superiorities including nominal performance recovery and chattering reduction. Both numerical and application examples have been simulated to demonstrate the effectiveness as well as the superiorities of the proposed method. The results have shown that the proposed method exhibits the properties of nominal performance recovery and chattering reduction as well as excellent dynamic and static performance as compared with the SMC and I-SMC methods.

Secondly, the continuous finite-time control problem of the system with mismatched disturbance has been addressed by using the nonsingular terminal sliding mode technique. A novel nonlinear dynamic sliding mode surface design has been proposed for the mismatched disturbance attenuation via a finite-time disturbance observer. The proposed method has the following two remarkable properties. First, the proposed method retains the nominal control performance since the FTDO serves like a patch to the baseline NTSMC and does not cause any adverse effects on the system in the absence of disturbance. Second, the proposed method largely alleviates the chattering problem of NTSMC since the mismatched disturbance has been compensated by FTDO-based compensation and no discontinuous control action is required to reject the disturbance.

APPLICATION TO PROCESS CONTROL SYSTEMS

Chapter 10

Disturbance Rejection for Typical Level Tank System

10.1 Introduction

Although DOBC has achieved plenty of elegant results in both control theory and applications, it mainly has its roots in mechanical engineering societies. In recent years, DOBC has become one of the hot topics in process control fields [1, 105, 133]. However, to the best of the authors knowledge, we can only find theoretical or simulation results of DOBC strategies in process control fields. Until now few experimental results have been reported in this fields, let alone industrial engineering applications. This chapter aims to develop a disturbance observer enhanced model predictive control method to improve the disturbance rejection ability of a typical industrial process control system with experimental results.

In addition, to achieve a faster response for the closed-loop system, a cascade composite control scheme is implemented to control a practical level tank that may be subject to severe disturbances. In such a composite control structure, the model predictive control serves as an outer-loop feedback controller, the PID control serves as an inner-loop feedback controller, and the DOB based control serves as a feed-forward compensator. Both the simulation and experimental studies have been carried out to show the effectiveness and feasibility of the proposed method.

10.2 System Modeling of Level Tank

Level tank control system is a typical and commonly used industry process, and the principal diagram of level tank studied here is shown in Figure 10.1. The controlled variable is the water level, and the manipulated variable is the input flow rate.

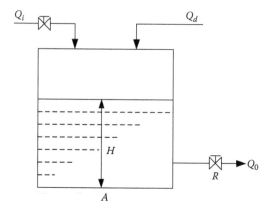

Figure 10.1   Abstract model of level tank system.

Obviously, the overall mass balance equation is

$$Q_i \rho + Q_d \rho - Q_o \rho = \rho \frac{dV}{dt}, \tag{10.1}$$

where Q_i the input flow rate of the level tank, Q_o the output flow rate of the level tank, Q_d the disturbance flow, V the capacity of the level tank and ρ the fluid density. Let H denotes the level height of the liquid inside the level tank, and A denotes the cross-sectional area of the level tank, we have

$$V = AH, \tag{10.2}$$

So Equation (10.1) can be rewritten as

$$Q_i + Q_d - Q_o = A \frac{dH}{dt}. \tag{10.3}$$

For such a level tank, the relationship between the output flow rate Q_o and the level height is

$$Q_o = \frac{H}{R}, \tag{10.4}$$

where R is the pipe resistance.

Substituting Equation (10.4) into Equation (10.3), we have

$$AR \frac{dH}{dt} + H = RQ_i + RQ_d. \tag{10.5}$$

It can be seen from Equation (10.5) that this is a first-order system. The time delay is generally unavoidable due to fluid transport delay through pipes or measurement sample delay. Making Laplace transformation for Equation (10.5) and taking into account the inherent time delay, the process channel model $G_p(s)$ can be represented as

$$G_p(s) = \frac{H(s)}{Q_i(s)} = \frac{Re^{-\theta s}}{ARs + 1}, \tag{10.6}$$

Let $AR = T$, $R = K$, Equation (10.6) becomes

$$G_p(s) = \frac{H(s)}{Q_i(s)} = \frac{Ke^{-\theta s}}{Ts + 1}. \tag{10.7}$$

10.3 Disturbance Rejection Control Design and Implementation

Since most of the process control systems can be modeled as first-order plus dead-time (FODT) or second-order plus dead-time (SODT) processes, the following model is considered to represent general process systems

$$Y(s) = G_p(s)U(s) + D_{ex}(s), \tag{10.8}$$

with

$$G_p(s) = g(s)e^{-\theta s}, \tag{10.9}$$

$$D_{ex}(s) = \sum_{i=1}^{m} G_{di}(s)D_i(s), \tag{10.10}$$

where $U(s)$ the manipulated variable, $Y(s)$ the controlled variable, $D_i(s)(i = 1, \ldots, m)$ the ith external disturbances, $D_{ex}(s)$ the effects of external disturbances on $Y(s)$, $G_p(s)$ the model of the process channel, $g(s)$ the minimum-phase part of $G_p(s)$, and $G_{di}(s)(i = 1, \ldots, m)$ the model of ith disturbance channel. The nominal model $G_n(s)$ can also be represented as a product of a minimum-phase part $g_n(s)$ and a dead-time part $e^{-\theta_n s}$, i.e.,

$$G_n(s) = g_n(s)e^{-\theta_n s}. \tag{10.11}$$

10.3.1 Model Predictive Control

The process dynamic of system (10.11) can be represented as

$$\hat{y}(t) = \sum_{k=1}^{\infty} A(k)\Delta u(t - k), \tag{10.12}$$

$$\Delta u(t) = u(t) - u(t - 1),$$

where $u(t)$ the manipulated variable, $\hat{y}(t)$ the output under the actions of $u(t-k)(k = 1, \ldots, \infty)$ and $A(k)$ is the dynamic matrix formed from the unit step response coefficients. According to (10.12), the ith step ahead prediction of the output with the prediction correction term is stated as

$$\hat{y}(t+i) = \sum_{k=1}^{i} A(k)\Delta u(t+i-k) + \sum_{k=1}^{\infty} A(k+i)\Delta u(t-k) + \xi(t),$$

$$\xi(t) = y(t) - \hat{y}(t), \qquad (10.13)$$

where $y(t)$ is the real time output, $\xi(t)$ represents the prediction correction term. In (10.13), $\Delta u(t+i-k)(k = 1, \ldots, i)$ denotes the future manipulated variable moves that are achieved by solving the following optimization problem

$$\min_{\Delta u(t)\ldots\Delta u(t+M-1)} J = \sum_{j=1}^{P} [e^T(t+j) W_e e(t+j)] + \sum_{j=0}^{M-1} [u^T(t+j) W_u u(t+j)],$$

$$e(t+j) = \hat{y}(t+j) - r(t+j), \qquad (10.14)$$

where P and M represent the prediction horizon and the control horizon, respectively, $e(t+j)$ is the prediction error, $r(t+j)$ is the desired reference trajectory, W_e and W_u represent the error weighting matrix and input weighting matrix, respectively. Only the first move is applied to the plant, and this step is repeated for the next sampling instance.

10.3.2 Disturbance Observer-Enhanced Model Predictive Control

The block diagram of the proposed disturbance observer enhanced model predictive control scheme is shown in Figure 10.2. $R(s)$ represents the reference trajectory of controlled variable. $C(s)$ stands for output of the MPC controller. $\hat{D}_f(s)$ denotes the disturbance estimation.

According to Figure 10.2, the output can be expressed as

$$Y(s) = G_{cy}(s)C(s) + G_{dy}(s)D_{ex}(s), \qquad (10.15)$$

where

$$G_{cy}(s) = \frac{g(s)e^{-\theta s}}{1 + Q(s)g_n^{-1}(s)[g(s)e^{-\theta s} - g_n(s)e^{-\theta_n s}]}, \qquad (10.16)$$

$$G_{dy}(s) = \frac{1 - Q(s)e^{-\theta_n s}}{1 + Q(s)g_n^{-1}(s)[g(s)e^{-\theta s} - g_n(s)e^{-\theta_n s}]}. \qquad (10.17)$$

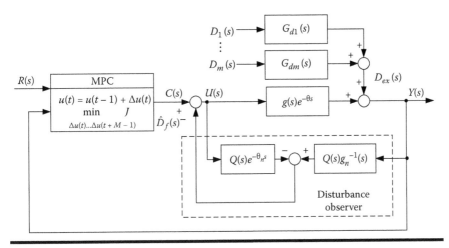

Figure 10.2   Block diagram of the disturbance observer-enhanced model predictive control.

It can follow from Equations (10.15)–(10.17) that the disturbance rejection property mainly depends on the design of filter $Q(s)$ in DOB. If we select $Q(s)$ as a low-pass filter with a steady-state gain of 1, i.e., $\lim_{\omega \to 0} Q(j\omega) = 1$, it can be obtained from (10.17) that $\lim_{\omega \to 0} G_{dy}(j\omega) = 0$. This means the low-frequency disturbances can be attenuated asymptotically. Here $Q(s)$ is selected as a first-order low-pass filter with a steady-state gain of 1, which can be expressed as

$$Q(s) = \frac{1}{\lambda s + 1}, \lambda > 0. \tag{10.18}$$

The robustness of the DOB is analyzed as follows. According to Figure 10.2, the output $Y(s)$ can be represented as

$$\begin{aligned} Y(s) &= g(s)e^{-\theta s}U(s) + D_{ex}(s) \\ &= g_n(s)e^{-\theta_n s}U(s) + D_m(s) + D_{ex}(s), \end{aligned} \tag{10.19}$$

where

$$D_m(s) = [g(s)e^{-\theta s} - g_n(s)e^{-\theta_n s}]U(s), \tag{10.20}$$

denotes the internal disturbances caused by model uncertainties.

Let the lumped disturbances consisting of external disturbances $D_{ex}(s)$ and internal disturbances $D_m(s)$

$$D_l(s) = D_m(s) + D_{ex}(s), \tag{10.21}$$

then, it follows from (10.19) that the output can be represented by

$$Y(s) = g_n(s)e^{-\theta_n s} U(s) + D_l(s).$$ (10.22)

As shown in Figure 10.2, the control law is

$$U(s) = C(s) - \hat{D}_f(s),$$ (10.23)

where

$$\hat{D}_f(s) = Q(s)g_n^{-1}(s)Y(s) - Q(s)e^{-\theta_n s} U(s),$$ (10.24)

Substituting Equation (10.22) into Equation (10.24) yields

$$\hat{D}_f(s) = Q(s)g_n^{-1}(s)D_l(s),$$ (10.25)

Define $\tilde{D}_l$ as the error between the real value and the estimated value of lumped disturbance, i.e.,

$$\tilde{D}_l(s) = D_l(s) - g_n(s)e^{-\theta_n s} \hat{D}_f(s)$$

$$= [1 - Q(s)e^{-\theta_n s}]D_l(s),$$ (10.26)

According to the final-value theorem, one obtains from Equation (10.26) that

$$\tilde{D}_l(\infty) = \lim_{t \to \infty} \tilde{D}_l(t)$$

$$= \lim_{s \to 0} s\, \tilde{D}_l(s)$$

$$= \lim_{s \to 0} [1 - Q(s)e^{-\theta_n s}] \lim_{s \to 0} s\, D_l(s)$$

$$= \lim_{s \to 0} [1 - Q(s)e^{-\theta_n s}] \lim_{t \to \infty} d_l(t)$$

$$= \lim_{s \to 0} [1 - Q(s)e^{-\theta_n s}]d_l(\infty),$$ (10.27)

Obviously, if we select the steady-state of $Q(s)$ to be 1, then one obtains from Equation (10.27) that

$$\tilde{D}_l(\infty) = 0.$$ (10.28)

Fine effects can be achieved by tuning the parameters of MPC controller and the time constants λ of filter $Q(s)$ in DOB. It should be pointed out that the implementation of DOB is rather simple, thus the introduction of feed-forward compensation part does not increase much computational complexity.

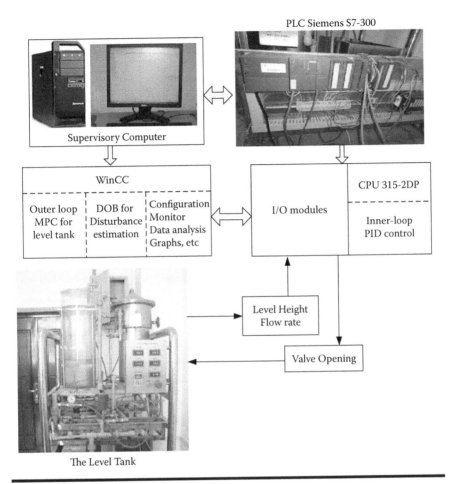

Figure 10.3   Hierarchy diagram of the composite level tank control system.

10.3.3 Control Implementation

In order to obtain a more prompt response, a local PID controller is employed to control the flow rate, which leads to a cascade control scheme. The experimental devices of level tank studied here are shown in Figure 10.3.

All the signals from or to the level tank system are intercollected through digital-to-analog or analog-to-digital interfaces in a supervisory control and data acquisition (SCADA) system, which works in a hierarchical manner. DOB-based MPC strategy is programmed by the configuration software WinCC from Siemens on a supervisory computer and final commands are carried out through the programmable logic controller (PLC). The control command from PLC will change the opening of valve from 0% to 100%. The detailed control structure is shown in Figure 10.4.

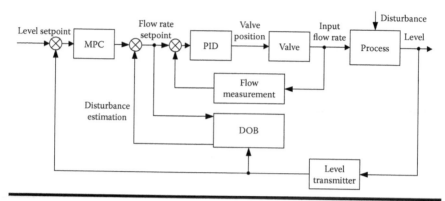

Figure 10.4   Control structure of the level tank system via the proposed method.

In this strategy, the disturbance observer-enhanced MPC is known as the primary or outer-loop controller, while the PID controller is the secondary or inner-loop controller.

From Figure 10.4, we can see that the disturbance directly affects the primary output (level). Note that the disturbance models are not necessarily required for the proposed method, thus only the process channel is modeled via step response test and represented as

$$G_n(s) = \frac{9.98}{11.965s + 1} e^{-1.8s}. \tag{10.29}$$

The nominal values of level tank system are: water level, 327 *mm* and input flow rate, 250 *L/h*.

According to the discussions above, the filter of DOB is selected as

$$Q(s) = \frac{1}{0.3s + 1}, \tag{10.30}$$

The MPC controller parameters of the proposed method for the level tank system are selected as

$$P = 8, \ M = 1, \ Ts = 1/6 \text{ min}, \ W_e = I, \ W_u = 1$$

The local PID controller parameters are designed as

$$K_p = 6.78, \ \tau_I = 2.6 \text{ min}$$

10.4   Simulation and Experimental Studies

To demonstrate the benefits of the proposed method, the baseline cascade control method consisting of an outer-loop MPC and an inner-loop PID controller is employed here for the purpose of comparison.

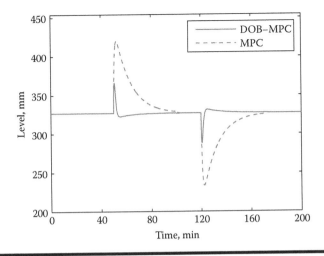

Figure 10.5 **Tank level response under the MPC (broken line) and the proposed method (solid line) (simulation).**

10.4.1 Simulation Results and Analysis

During the simulation studies, it is supposed that the disturbance (e.g., another input flow rate variation) is imposed on the input channel. Actually, the disturbance appears as

$$d(t) = \begin{cases} 0 \text{ L/h} & \text{for } 0 \leq t < 50, \\ 150 \text{ L/h} & \text{for } 50 \leq t < 120, \\ 0 \text{ L/h} & \text{for } t \geq 120. \end{cases} \tag{10.31}$$

In this case, the tank level and input flow rate under the control of both MPC and the proposed methods are shown in Figures 10.5 and 10.6, respectively.

It can be observed from Figures 10.5 and 10.6 that the proposed method significantly improves the disturbance rejection performance as compared with the baseline MPC-based cascade control method, including a much smaller overshoot and a shorter settling time.

10.4.2 Experimental Results and Analysis

To evaluate the practicality of the proposed method, experimental studies are also implemented. The disturbance added here is different but quite similar to that of the simulation case. In fact, another input flow rate variation is imposed on the system by setting the valve from close to open, and then close finally. Since the flow rate variation is large, it is supposed that strong disturbances occur here.

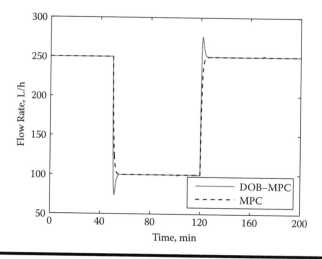

Figure 10.6 Input flow rate under the MPC (broken line) and the proposed method (solid line) (simulation).

The experimental data of the tank level and the input flow rate under both the MPC and the proposed methods are shown in Figures 10.7 and 10.8, respectively.

As shown in Figures 10.7 and 10.8, the overshoot under the proposed method is 28%, while it is 33% under the MPC method. The settling time under the proposed method is about 190 seconds, while it is 540 seconds under the MPC method. In

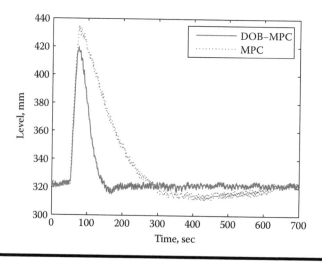

Figure 10.7 Tank level response under the MPC (broken line) and the proposed method (solid line) (experiment).

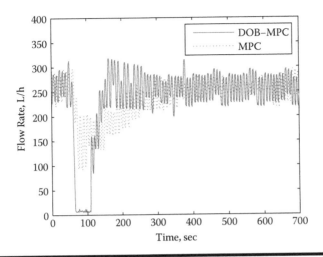

Figure 10.8 Input flow rate under the MPC (broken line) and the proposed method (solid line) (experiment).

general, the proposed method achieves much better disturbance rejection property than that of the MPC method in both steady state and transient state. To this end, it can be said that the proposed method provides an effective way in improving disturbance rejection performance for such kind of process control systems.

10.5 Summary

A composite cascade anti-disturbance control consisting of model predictive control, PID control, and disturbance observer has been proposed to control the first-order plus time-delay process. The MPC has advantages in handling dead-time processes due to its prediction mechanism. DOB has been used to estimate the disturbance and the estimation has been used for feed-forward compensation design to reject disturbance. Such a disturbance observer-enhanced model predictive control method has been applied to control the level tank which may suffer from strong disturbances. Both simulation and experimental studies have been carried out, and the results have demonstrated that the proposed method exhibits excellent disturbance rejection performance, such as a smaller overshoot and a shorter settling time.

Chapter 11

Disturbance Rejection for Ball Mill Grinding Circuits

11.1 Introduction

Ball mill grinding circuits are amongst the most important operation units in mineral processing plants. Grinding processes hold almost 50% of the total expenditure of concentrator plant. Moreover, the product particle size influences the recovery rate of valuable minerals as well as the volume of tailing discharge in the subsequent processes. Generally speaking, the aim of grinding processes is to maintain stable and effective control of the product particle size and the circulating load. However, undesirable characteristics, such as strong interactions between the variables, complex dynamic features, time delays, and strong disturbances, always bring about difficulties and problems to both control engineers and researchers. To this end, besides traditional PID controllers [134], many advanced approaches, such as model predictive control (MPC) [134, 135, 136, 137, 138], adaptive control [22, 139], neuro-control [140, 141], robust control [21] and supervisory expert control [142, 143], etc., are reported in recent years.

As mentioned above, MPC has been researched and applied as one of the most popular advanced control schemes in grinding field. The reason may lie in that MPC possesses the following remarkable advantages: (1) capability of handling multi-input-multi-output (MIMO) processes, (2) ability to control difficult processes, such as processes with time delay part and non-minimum phase part, (3) capability of handling constraints of both manipulated variables and controlled variables, and (4) simplicity of modeling (generally depending on step/impulse response), etc.

However, in ball mill grinding circuit, the control performance of closed-loop system is severely interrupted by various disturbances which are generally complex

171

and also unmeasurable. Usually, these disturbances can be divided into external and internal ones. The external disturbances often include the variations of ore hardness and feed particle size, and so on, while the internal disturbances are generally caused by model mismatches or coupling effects. Variations of ore hardness and feed particle size may cause continuous fluctuation of product particle size. Model mismatches and coupling effects may lead to adverse dynamic performance and even result in unstable control of the closed-loop system.

It should be pointed out that MPC, as well as many other advanced control schemes, rejects disturbances merely through the action of feedback regulation part and does not deal with disturbances directly by controller design. When meeting strong disturbances, these methods may lead to some limitations, for example, the dynamic process of closed-loop system may become sluggish and the closed-loop system may even become unstable due to the reason that the fluctuations of the controlled variables may become too large.

In order to improve the disturbance rejection performance of grinding circuits, a feedforward compensation part for disturbances is introduced to the controller besides a conventional feedback part. However, usually, it is hard or even impossible to measure the disturbances of grinding circuits. A feasible solution is to develop disturbance estimation technique. Disturbance observer (DO) is known as an effective technique to estimate disturbances and has been extensively applied for feedforward compensation design in the presence of disturbances.

In this chapter, a composite control scheme combining a feedforward compensation part based on DOs and a feedback regulation part based on MPC (DOB-MPC) is developed to improve the disturbance rejection performance of the closed-loop grinding system. This chapter mainly refers to [1]. One novelty of this work lies in that the terms of model mismatches in the closed-loop system are merged into the disturbance term and are regarded as a part of the lumped disturbances, in other words, the disturbance rejection analysis here considers not only the external disturbances, but also the effects caused by model mismatches. The other novelty lies in that a multivariable compound control structure based on DO and MPC is proposed, thus the DOB-MPC scheme inherits many advantages of both MPC and DO, some of which are listed as follows:

- Capability in handling processes with time delays;
- Ability to dealing with multivariable systems (good decoupling property);
- Remarkable superiorities in handling strong disturbances (including both external disturbances and internal disturbances caused by model mismatches).

11.2 Process Description

11.2.1 Process Background

Raw ores can not be used as final products for industrial uses. The aim of mineral processing is to concentrate the valuable minerals for the subsequent extraction processing [144]. The valuable minerals are originally contained in an oxide or sulfide. It is a long and difficult process to liberate the valuable minerals from the coarse ores. Usually, the overall mineral processing can be divided into the following three parts:

■ Ore size reduction process (crushing, grinding and classifying);
■ Separation (flotation, magnetic or gravimetric separation) and concentration process;
■ Tailing discharge.

The raw ores can not be sent to ball mill directly because the size is generally too large for grinding. Crushing process, aiming to reduce the size of raw ore, supplies appropriate size of feed ore for grinding process. The quality of product particle will affect the subsequent separation process. Moreover, the efficiency of this process has tight relation with the overall economic benefits of concentrator plant. According to the ore property, specific methods can be chosen for separation, such as flotation, magnetic and gravimetric separation, etc. Finally, tailings disposal is of great importance from the viewpoint of environment protection.

11.2.2 Description of Grinding Circuit

Ball mill grinding circuit is one of the most important mineral processing units in concentrator plants. A typical ball mill grinding circuit is shown in Figure 11.1, which consists of a ball mill, a pump sump, hydro-cyclones and solids feeding conveyors.

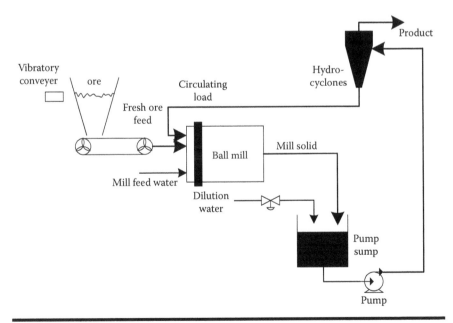

Figure 11.1 Schematic diagram of a ball mill grinding circuit.

When the ball mill works, the fresh ore is fed into the mill together with water. The tumbling action of balls within the revolving mill crushes the feed ore to finer sizes. The slurry containing the fine product is discharged from the mill to a pump sump, and then pumped to hydro-cyclones for classification. After the classification process, the slurry is separated into two streams: an overflow stream containing the finer particles and an underflow stream containing the larger particles as the circulating load. The finer particles are the desired product, while the circulating load is recycled back to the ball mill for regrinding.

Product particle size y_1 (%—$200mesh$) is the most important controlled variable in a grinding circuit. Mine ore has to be ground to a specified size called the liberation size that the imbedded mineral particles are exposed for effective recovery in the downstream processing. Circulating load y_2 (t/h) is another important controlled variable. Larger or smaller circulating load than the setpoint may degrade the efficiency of the whole circuit. Note that mill solids concentration (%$solids$) and sump level (m) are another two important variables need to be controlled. In real practice, two local controllers are usually designed for such two controlled variables: the sump level is maintained to be constant by regulating the slurry pumping rate (Hz), while the mill solids concentration is kept around a desired setpoint by regulating the mill addition water (m^3/h) in (constant or variable) proportion to the fresh ore feed rate [136, 142]. Thus, only two controlled variables, viz product particle size y_1 and circulating load y_2 are considered.

Strong external disturbances in grinding circuits, such as variations of ore hardness d_{oh} and feed particle size d_{fps}, may cause continuous fluctuations of product particle size. Both fresh ore feed rate u_1 (t/h) and dilution water flow rate u_2 (m^3/h) can be manipulated to regulate the two selected controlled variables [142]. The relationships among the selected controlled variables, manipulated variables, and external disturbances in grinding process are illustrated clearly in Figure 11.2.

11.3 Control Scheme

In this work, a composite control scheme is proposed to solve difficult control problems in grinding circuit. Firstly, multivariable MPC algorithm and DO technique are introduced. A DOB-MPC control scheme is then illustrated in details.

11.3.1 Multivariable MPC Algorithm

Multivariable MPC algorithm was originally proposed by Cutler et al. [145] and discussed extensively by many researchers. Here it is summarized for the convenience of readers. Considering a multivariable system with m manipulated variables and

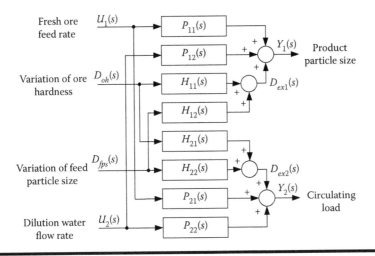

Figure 11.2   Relationships among the variables in grinding circuit.

n controlled variables, the process dynamics can be represented as

$$\hat{y}(t) = \sum_{k=1}^{\infty} A(k)\Delta u(t - k),$$

$$\Delta u(t) = u(t) - u(t - 1), \tag{11.1}$$

where $u(t) \in R^m$ is the vector of the manipulated variable, $\hat{y}(t) \in R^n$ is the output vector under the actions of $u(t - k)(k = 1, \ldots, \infty)$ and $A(k)$, with dimensions of $n \times m$, is the dynamic matrix formed from the unit step response coefficients. According to (11.1), the ith step ahead prediction of the output with the prediction correction term is stated as

$$\hat{y}(t + i) = \sum_{k=1}^{i} A(k)\Delta u(t + i - k) + \sum_{k=1}^{\infty} A(k + i)\Delta u(t - k) + d(t),$$

$$d(t) = y(t) - \hat{y}(t), \tag{11.2}$$

where $y(t)$ is the vector of the realtime output, $d(t)$ represents the prediction correction term and consists of the measurement noise, the external disturbances as well as the effects caused by model mismatches. In (11.2), $\Delta u(t + i - k)(k = 1, \ldots, i)$ denote the future manipulated variable moves and are obtained by solving an optimization problem. A general quadratic performance objective function of the MPC optimization problem possesses the following form

$$\min_{\Delta u(t)\cdots\Delta u(t+M-1)} J = \sum_{j=1}^{P} [e^{T}(t+j)\, Qe(t+j)] + \sum_{j=0}^{M-1} [u^{T}(t+j)\, Ru(t+j)],$$
$$e(t+j) = \hat{y}(t+j) - r(t+j), \tag{11.3}$$

subject to the following constraints:

$$u^{L} \leq u(t+j) \leq u^{H} : \text{input saturation limits,}$$
$$y^{L} \leq y(t+j) \leq y^{H} : \text{output specification limits,} \tag{11.4}$$
$$\Delta u(t+j) = 0 \text{ for } j > M,$$

where P and M represent the prediction horizon and the control horizon, respectively, $e(t+j)$ is the prediction error, $r(t+j)$ is the desired reference trajectory, Q and R represent the error weighting matrix and input weighting matrix, respectively.

MPC scheme is implemented in a receding horizon framework. At any sampling instance, the optimization problem is formulated over the prediction horizon and an optimized manipulated variable trajectory is solved from the objective function. Only the first move is applied to the plant and this step is repeated for the next sampling instance.

11.3.2 Disturbance Observer for Process with Time Delays

In grinding process, the conventional disturbance observer is no longer available due to the reason that the inverse function of time delay part is physically unrealizable. To this end, a modified disturbance observer with considerations of time delays [105] is adopted here. Consider a two-input-two-output process with two external disturbances which is expressed as

$$\begin{bmatrix} Y_1(s) \\ Y_2(s) \end{bmatrix} = \begin{bmatrix} P_{11}(s) & P_{12}(s) \\ P_{21}(s) & P_{22}(s) \end{bmatrix} \begin{bmatrix} U_1(s) \\ U_2(s) \end{bmatrix} + \begin{bmatrix} H_{11}(s) & H_{12}(s) \\ H_{21}(s) & H_{22}(s) \end{bmatrix} \begin{bmatrix} D_{oh}(s) \\ D_{fps}(s) \end{bmatrix}, \tag{11.5}$$

with

$$P_{ii}(s) = p_{ii}(s)e^{-\tau_{ii}s} (i = 1, 2), \tag{11.6}$$
$$D_{exi}(s) = H_{i1}(s)D_{oh}(s) + H_{i2}(s)D_{fps}(s)(i = 1, 2), \tag{11.7}$$

where $U_i(s)(i = 1, 2)$ the manipulated variables, $Y_i(s)(i = 1, 2)$ the controlled variables, $D_{oh}(s)$ and $D_{fps}(s)$ the external disturbances, $D_{exi}(s)(i = 1, 2)$ the effects of external disturbances on $Y_i(s)$, $P_{ij}(s)(i = 1, 2; j = 1, 2)$ the models of the process channels, $p_{ii}(s)(i = 1, 2)$ the minimum-phase part of $P_{ii}(s)$, and $H_{ij}(s)(i = 1, 2; \ j = 1, 2)$ the models of disturbance channels.

Assume that u_1 is manipulated to control y_1, also u_2 is manipulated to control y_2. The structure of modified disturbance observer for $u_1 - y_1$ loop is shown in Figure 11.3(a). Note that the case of $u_2 - y_2$ loop is similar which is omitted here. Clearly, model $P_{11}(s)$ is separated into a multiplication of a minimum-phase part p_{11} and a time delay part $e^{-\tau_{11}s}$ as below

$$P_{11}(s) = p_{11}(s)e^{-\tau_{11}s}. \tag{11.8}$$

Similarly, the nominal model of the $u_1 - y_1$ loop, i.e., $G_{11}(s)$ can be represented as

$$G_{11}(s) = g_{11}(s)e^{-\tau_{11n}s}. \tag{11.9}$$

Considering the effect of model mismatches, an equivalent block diagram is obtained, as shown in Figure 11.3(b). The final form of modified disturbance observer is shown in Figure 11.3(c). Clearly, lumped disturbance $D_1(s)$ of $u_1 - y_1$ loop includes external disturbance $D_{ex1}(s)$ and internal disturbances consisting of $D_{m1}(s)$ caused by model mismatches and $D_{c1}(s)$ caused by coupling effects, that is,

$$D_1(s) = D_{ex1}(s) + D_{m1}(s) + D_{c1}(s), \tag{11.10}$$

where

$$D_{m1}(s) = [p_{11}(s)e^{-\tau_{11}s} - g_{11}(s)e^{-\tau_{11n}s}]U_1(s), \tag{11.11}$$

$$D_{c1}(s) = p_{12}(s)e^{-\tau_{12}s}U_2(s). \tag{11.12}$$

One obtains from Figure 11.3(c) that

$$Y_1(s) = g_{11}(s)e^{-\tau_{11n}s}U_1(s) + D_{m1}(s) + D_{ex1}(s) + D_{c1}(s). \tag{11.13}$$

Substituting (11.10) into (11.13), yields

$$Y_1(s) = g_{11}(s)e^{-\tau_{11n}s}U_1(s) + D_1(s). \tag{11.14}$$

It can be obtained from Figure 11.3(c) that

$$\hat{D}_{f1}(s) = Q(s)g_{11}^{-1}(s)Y_1(s) - Q(s)e^{-\tau_{11n}s}U_1(s). \tag{11.15}$$

Substituting (11.14) into (11.15), yields

$$\hat{D}_{f1}(s) = Q(s)g_{11}^{-1}(s)D_1(s). \tag{11.16}$$

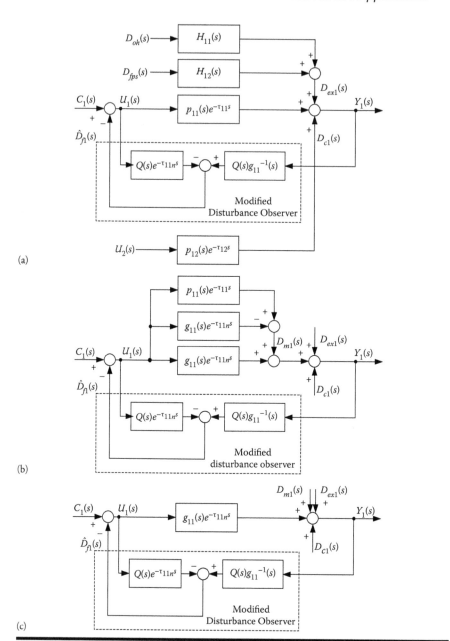

Figure 11.3 Block diagram of the modified disturbance observer for the process with time delays: (a) an initial form, (b) an equivalent form, (c) the final form.

Define $\bar{D}_i(s)(i = 1, 2)$ as the error between the real value and the estimation value of lumped disturbance in $u_i - y_i$ loop, that is,

$$\bar{D}_1(s) = D_1(s) - g_{11}(s)e^{-\tau_{11n}s} \hat{D}_{f1}(s). \tag{11.17}$$

Substituting (11.16) into (11.17), yields

$$\bar{D}_1(s) = [1 - Q(s)e^{-\tau_{11n}s}] D_1(s). \tag{11.18}$$

According to the final-value theorem, one obtains from (11.18) that

$$\begin{aligned}
\bar{d}_1(\infty) &= \lim_{t \to \infty} \bar{d}_1(t) \\
&= \lim_{s \to 0} s \, \bar{D}_1(s) \\
&= \lim_{s \to 0} [1 - Q(s)e^{-\tau_{11n}s}] \lim_{s \to 0} s \, D_1(s) \\
&= \lim_{s \to 0} [1 - Q(s)e^{-\tau_{11n}s}] \lim_{t \to \infty} d_1(t) \\
&= \lim_{s \to 0} [1 - Q(s)e^{-\tau_{11n}s}] d_1(\infty). \tag{11.19}
\end{aligned}$$

Obviously, if selecting $Q(s)$ such that the steady-state gain of $Q(s)$ is 1, then it can be obtained that

$$\bar{d}_1(\infty) = 0. \tag{11.20}$$

It can be found from (11.20) that the disturbances can be asymptotically rejected. Hence, the modified disturbance observer is a practical approach to deal with the process with time delays.

11.3.3 DOB-MPC Scheme for Ball Mill Grinding Circuits

The method of this work focuses on disturbance rejection against coupling effects, model mismatches as well as external disturbances. A DOB-MPC scheme is proposed to control the ball mill grinding circuits in this part. The detailed design procedures are described as follows.

Consider a typical two-input-two-output grinding circuit shown in Figure 11.1. The manipulated variables are fresh ore feed rate u_1 (t/h) and dilution water flow rate u_2 (m^3/h), while the controlled variables are product particle size y_1 ($\% - 200mesh$) and circulating load y_2 (t/h), respectively. According to the requirements of the real practice, the constraints of manipulated and controlled variables are stated as

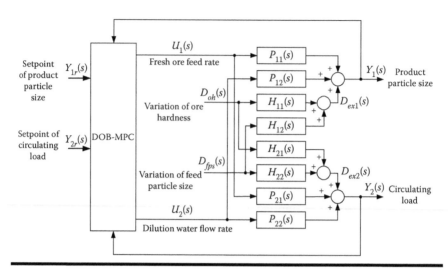

Figure 11.4 Schematic block diagram of the feedforward-feedback DOB-MPC scheme for grinding circuit.

follows: (1) the minimum and maximum values of fresh ore feed rate are 60 (t/h) and 70 (t/h), (2) the minimum and maximum values of the dilution water flow rate are 40 (m^3/h) and 50 (m^3/h), (3) the minimum and maximum values of the product particle size are 68 $(\% - 200\,mesh)$ and 72 $(\% - 200\,mesh)$, and (4) the minimum and maximum values of circulating load are 140 (t/h) and 170 (t/h). Assume that fresh ore feed rate u_1 is manipulated to control the product particle size y_1, and dilution water flow rate u_2 is manipulated to control the circulating load y_2. In this case, a schematic block diagram of the DOB-MPC scheme for the ball mill grinding circuit is shown in Figure 11.4.

The internal structure of the feedforward-feedback DOB-MPC controller is shown in Figure 11.5. It can be observed that a MPC controller and two disturbance observers are employed here to construct a composite DOB-MPC structure. Fine effects can be achieved by tuning the parameters of MPC controller and the time constants of filter $Q(s)$ in DO. Here $Q(s)$ is selected as a first-order low-pass filter with a steady-state gain of 1, which can be expressed as

$$Q(s) = \frac{1}{\lambda s + 1}, \quad \lambda > 0. \tag{11.21}$$

It should be pointed out that the implementation of DO is rather simple, thus the introduction of feedforward compensation part does not increase much computational complexity.

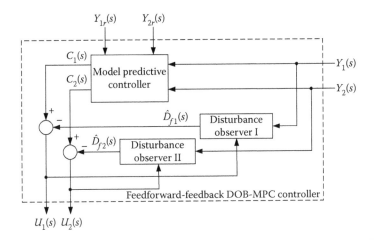

Figure 11.5   Internal structure of the feedforward-feedback DOB-MPC controller.

11.4 Performance Analysis and Comparisons

For control study, plant test has been constructed to develop the input-output transfer function model of grinding circuit as follows:

$$\begin{bmatrix} Y_1(s) \\ Y_2(s) \end{bmatrix} = \begin{bmatrix} G_{11}(s) & G_{12}(s) \\ G_{21}(s) & G_{22}(s) \end{bmatrix} \begin{bmatrix} U_1(s) \\ U_2(s) \end{bmatrix}, \tag{11.22}$$

where

$$G_{11}(s) = g_{11}(s)e^{-\tau_{11n}s} = \frac{-0.58}{2.5s+1}e^{-0.68s}, \tag{11.23}$$

$$G_{12}(s) = g_{12}(s)e^{-\tau_{12n}s} = \frac{4(1-0.9938e^{-0.47s})}{(2s+1)(6s+1)}e^{-0.2s}, \tag{11.24}$$

$$G_{21}(s) = g_{21}(s)e^{-\tau_{21n}s} = \frac{2.2}{6s+1}e^{-0.6s}, \tag{11.25}$$

$$G_{22}(s) = g_{22}(s)e^{-\tau_{22n}s} = \frac{2.83}{3.5s+1}e^{-0.13s}. \tag{11.26}$$

Note that the time constants are expressed in minutes here. As for such nominal model, a feasible variable pairing is $u_1 - y_1$ and $u_2 - y_2$ according to relative gain array (RGA) analysis. In this section, the disturbance rejection performance is studied in the nominal case as well as the model mismatch case. Here, the external disturbances are imposed on the process through disturbance channels. According to Reference [20], ore hardness and feed particle size have great influences on the product particle size and the circulating load. For example, ore feed with increased ore hardness leads to a decreased particle size and an increased circulating load, and these dynamics can be modeled as a first-order plus dead-time (FOPDT) form, which is the most commonly used model to describe the dynamics of industrial process. The grinding process can be described as (11.5). For simulation study, consider that the transfer functions of disturbance channels $H_{ij}(s)$ are obtained by mechanism analysis of process and expressed as follows

$$H_{11}(s) = \frac{-6.4}{4.8s+1} e^{-0.52s}, \tag{11.27}$$

$$H_{12}(s) = \frac{-8.6}{5.2s+1} e^{-0.58s}, \tag{11.28}$$

$$H_{21}(s) = \frac{32}{5.3s+1} e^{-0.46s}, \tag{11.29}$$

$$H_{22}(s) = \frac{40}{6.2s+1} e^{-0.63s}, \tag{11.30}$$

In (11.5), $D_{oh}(s)$ and $D_{fps}(s)$ represent variations of ore hardness and feed particle size, respectively. It should be pointed out that both ore hardness and feed particle size are hard or impossible to be measured. Thus, here they are expressed in a relative change form rather than a real physical unit form. For example, $D_{oh}(s) = 10\%$ means that the ore hardness has an increase of 10% compared with its nominal value.

The disturbance rejection performance under the control of proposed method is compared with that under the MPC method. The tuning parameters of the two controllers are listed in Table 11.1.

11.4.1 Disturbance Rejection in Nominal Case

Since here we study the external disturbance rejection performance of grinding process under the control of the proposed method, the nominal case is considered, i.e., the transfer function models of process channels satisfy that $P_{ij}(s) = G_{ij}(s)(i =$

Table 11.1 Tuning Parameters of the Controllers for Grinding Process

Method	$T_s(min)$	P	M	Q	R	$\lambda_1(min)$	$\lambda_2(min)$
MPC	0.5	40	10	diag[1.0 0.4]	diag[0.5 0.5]	---	---
DOB-MPC	0.5	40	10	diag[0.2 0.2]	diag[1.0 1.0]	0.1	1.5

1, 2; $j = 1, 2$). In this part, two kinds of simulations have been carried out to demonstrate the external disturbance rejection property of the proposed method. Generally, the ore hardness and feed particle size fluctuates continuously during the whole process of production, thus a sinusoidal disturbance is likely to match the features of the real practice better than a common step disturbance does. To this end, besides the step disturbance case, a sinusoidal disturbance case is also considered and simulated.

Case I: Step external disturbances in the nominal case: the ore hardness has an increase of 20% at $t = 20$ *min*, while the feed particle size has an decrease of 20% at $t = 100$ *min*.

Figure 11.6 shows the response curves of controlled variables under the control of both DOB-MPC (denoted by solid line) and MPC (denoted by dash-dot line) in *Case I*. The corresponding changes of manipulated variables are shown in Figure 11.7. The effects of external disturbances and the estimations on the controlled variables are shown in Figure 11.8. It can be observed from Figure 11.6 that the dynamic performances of both product particle size and circulating load under the proposed method (solid line) are much better than those under the MPC method (dash-dot line). Compared with the MPC scheme, the proposed method obtains a faster convergence speed, smaller amplitudes of fluctuations and shorter settling times, etc. As shown in Figure 11.8, the errors between the estimated (solid line) and real (dot line) disturbances are very small, which means that the disturbance observer can effectively estimate the effects caused by disturbances.

To quantitatively analyze the disturbance rejection performance, two performance indexes including peak overshoot and integral of absolute error (IAE) are employed. The performance indexes in such case are shown in Table 11.2. It can be observed from Table 11.2 that the overshoots for both product particle size and circulating load under the proposed method are much smaller than those under the MPC method. The IAE values of both controlled variables under the proposed method are also much smaller that those under the MPC scheme.

Case II: Sinusoidal external disturbances in the nominal case: variation of the ore hardness is sinusoidal, i.e., $d_{oh} = A\sin(\omega t)$, here $A = 20\%$, $\omega = 0.1$ *rad/min*.

Figure 11.9 shows the response curves of the controlled variables under the control of the two methods in the case of a sinusoidal disturbance. The corresponding variations of manipulated variables are shown in Figure 11.10. Also the effects of real disturbances and the estimations by disturbance observer are shown in Figure 11.11.

It can be observed from Figure 11.9 that the fluctuation amplitudes of both product particle size and circulation load under the proposed method (solid line) are much smaller than those under the MPC method (dash-dot line), although the fluctuation frequencies of both two controlled variables under the two control methods are almost the same. Figure 11.10 shows that both the fluctuation amplitudes and frequencies of the two manipulated variables under the proposed method (solid line) are almost the same as those under the MPC scheme (dash-dot line). However, it can be observed from Figure 11.10 that variations of the manipulated variables

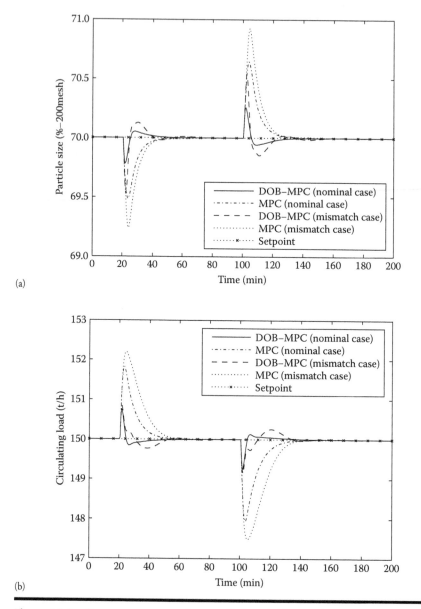

(a)

(b)

Figure 11.6 **Response curves of controlled variables in the presence of step external disturbances under DOB-MPC and MPC schemes in both nominal case and model mismatch case: (a) particle size, (b) circulating load.**

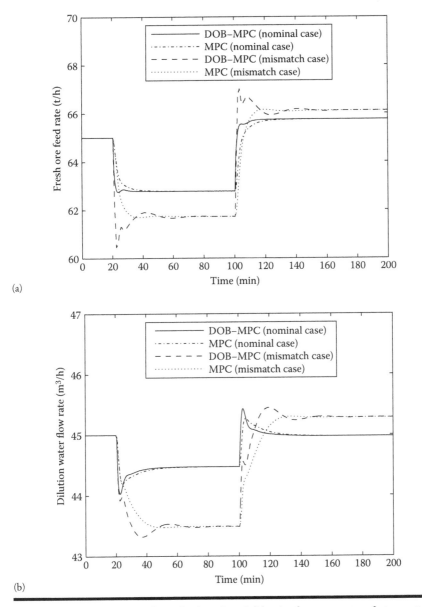

(a)

(b)

Figure 11.7 **Variations of manipulated variables in the presence of step external disturbances under DOB-MPC and MPC schemes in both nominal case and model mismatch case: (a) fresh ore feed rate, (b) dilution water flo w rate.**

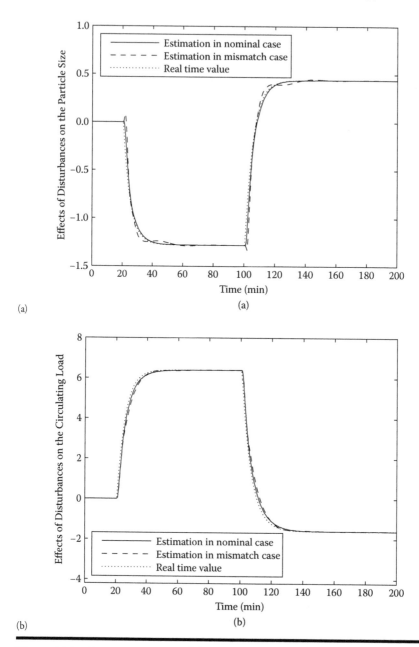

(a)

(b)

Figure 11.8 **Response curves of disturbances and their estimations in the presence of step external disturbances under the DOB-MPC scheme in both the nominal case (solid line) and model mismatch case (dashed line): (a) effects of the disturbances on the particle size, (b) effects of the disturbances on the circulating load. The real disturbances are denoted by a dotted line.**

Table 11.2 Performance Indexes in the Presence of Step External Distur-bances for the Nominal Case

Performance index	Particle size (DOB-MPC)	Particle size (MPC)	Circulating load (DOB-MPC)	Circulating load (MPC)
Overshoot(%)	0.37	0.92	0.50	1.37
IAE	2.77	9.80	7.12	37.97

under the proposed method are a little quicker than those under the MPC method. Figure 11.11 shows that the estimated (solid line) and the real (dotted line) distur-bances almost overlap. Thus it can be said that the disturbance observer can estimate the disturbances effectively and the proposed method is more effective to overcome such sinusoidal external disturbance than the MPC scheme.

11.4.2 Disturbance Rejection in Model Mismatch Case

In real practice, besides external disturbances, internal disturbances caused by model mismatches is another important factor which affects the control performance of the closed-loop systems. As illustrated in Section 11.3, the proposed method can reject not only external disturbances, but also the internal disturbance caused by model mismatches. In this part, we aim to demonstrate the lumped disturbance rejection performance of the proposed method through simulation studies.

Suppose that the transfer function models of process channels are expressed as

$$P_{11}(s) = \frac{-0.4}{3.5s + 1} e^{-0.5s}, \tag{11.31}$$

$$P_{12}(s) = \frac{3.42(1 - 0.994e^{-0.47s})}{(s + 1)(2s + 1)} e^{-0.13s}, \tag{11.32}$$

$$P_{21}(s) = \frac{0.8}{3.2s + 1} e^{-0.3s}, \tag{11.33}$$

$$P_{22}(s) = \frac{2.5}{5.5s + 1} e^{-0.2s}. \tag{11.34}$$

Comparing (11.23)-(11.26) with (11.31)-(11.34), it can be found that severe model mismatches exist.

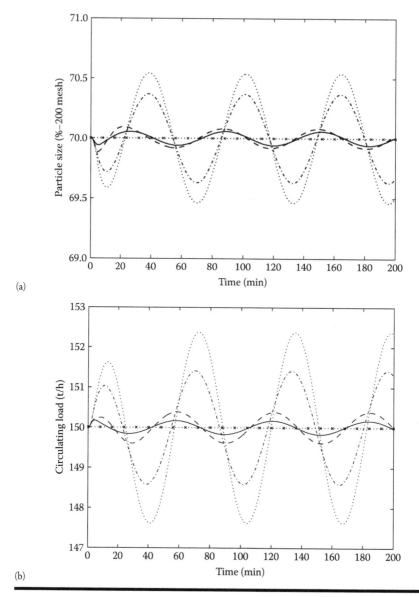

(a)

(b)

Figure 11.9 **Response curves of controlled variables in the presence of sinusoidal external disturbances under DOB-MPC (solid line for nominal case and dashed line for model mismatch case) and MPC (dash-dot line for nominal case and dotted line for model mismatch case) schemes in both nominal case and model mismatch case: (a) particle size, (b) circulating load. The setpoints are denoted by x-marked dotted line.**

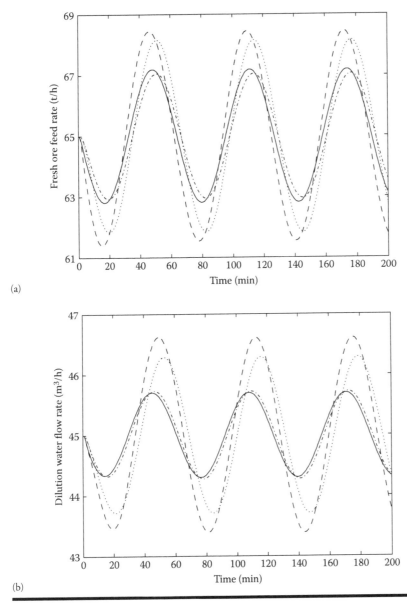

(a)

(b)

Figure 11.10 **Variations of manipulated variables in the presence of sinusoidal external disturbances under DOB-MPC (solid line for nominal case and dashed line for model mismatch case) and MPC (dash-dot line for nominal case and dotted line for model mismatch case) schemes in both nominal case and model mismatch case: (a) fresh ore feed rate, (b) dilution water flow rate.**

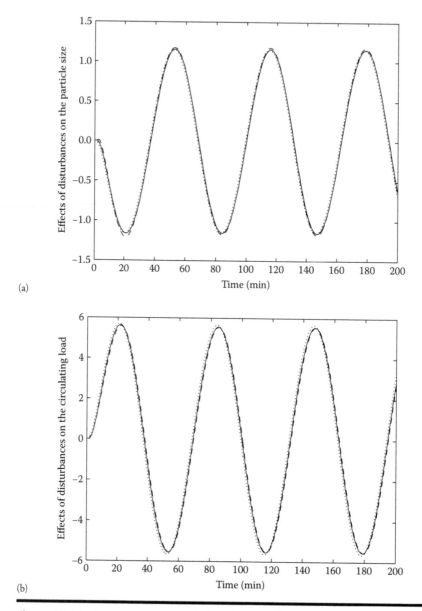

(a)

(b)

Figure 11.11 Response curves of disturbances and their estimations in the presence of sinusoidal external disturbances under DOB-MPC scheme in both nominal case (solid line) and model mismatch case (dashed line): (a) effects of the disturbances on the particle size, (b) effects of the disturbances on the circulating load. The real disturbances are denoted by dotted line.

Table 11.3 Performance Indexes in the Presence of Step External Distur-bances for the Model Mismatch Case

Performance index	Particle Size (DOB-MPC)	Particle Size (MPC)	Circulating Load (DOB-MPC)	Circulating Load (MPC)
Overshoot(%)	0.77	1.32	0.57	1.68
IAE	5.99	14.26	13.82	68.45

Case III: Step external disturbances in the model mismatch case: the ore hardness has an increase of 20% at $t = 20$ *min*, while the feed particle size has an decrease of 20% at $t = 100$ *min*.

The response curves of the controlled variables under the control of the two methods (dashed line for DOB-MPC and dotted line for MPC) in *Case III* are shown in Figure 11.6. The corresponding response curves of manipulated variables are show in Figure 11.7. Estimations of disturbances in such case (denoted by dashed line) are shown in Figure 11.8. Table 11.3 lists the performance indexes in such case.

Similar with *Case I*, it can be observed from Figure 11.6 and Table 11.3 that the proposed method possesses a smaller peak overshoot, a smaller values of IAE and a faster convergence speed, etc. This means that the proposed method has achieved a much better step disturbance rejection performance than that of the MPC method even in the case of severe model mismatches. Figure 11.8 shows that, even in such severe model mismatches, the errors between the estimated and the real disturbances are also very small.

Case IV: Sinusoidal external disturbances in the model mismatch case: variation of ore hardness is sinusoidal, i.e., $d_{oh} = A\sin(\omega t)$, here $A = 20\%$, $\omega = 0.1$ *rad/min*.

In *Case IV*, the response curves of controlled variables under the control of the proposed method (dashed line) and the MPC method (dotted line) are shown in Figure 11.9. The corresponding changes of manipulated variables are shown in Figure 11.10. Figure 11.11 shows the effects of estimated disturbances in such case (dashed line).

It can be seen from Figure 11.9 that the fluctuation amplitudes of both product particle size and circulating load under the control of the proposed method (dashed line) are much smaller than those under the MPC method (dotted line). The fluctuation frequencies between the two methods are almost the same. Figure 11.10 shows that the fluctuation amplitudes of manipulated variables under the proposed method (dashed line) are a little larger than those under the MPC scheme (dotted line), but their changes are prior to those under the MPC method. It can be observed from Figure 11.11 that the disturbance estimation errors in *Case IV* (difference between dashed line and dotted line) are also very small. Simulation results of *Case IV* demonstrate that the proposed method has remarkable superiorities in rejecting such lumped disturbances consisting of sinusoidal external disturbances and internal disturbances caused by model mismatches.

11.5 Summary

In grinding process, complex and unmeasurable disturbances generally have undesirable influences on closed-loop system. Many existing methods including MPC demonstrate some limitations in the presence of strong disturbances. To improve the disturbance rejection property, a technique called disturbance observer (DO) has been introduced for feedforward compensation design in this book. A composite control scheme consisting of a feedforward compensation part based on DO and a feedback regulation part based on MPC has been developed. Besides external disturbances, internal disturbances caused by model mismatches are merged into the terms of disturbance and regarded as a part of the lumped disturbance. Rigorous analysis on disturbance rejection property have also been given with the considerations of both model mismatches and external disturbances. Simulation results show that, compared with the control effects of MPC method, the proposed method has obtained remarkable superiority in rejecting the lumped disturbances in grinding circuits.

APPLICATION TO MECHATRONIC SYSTEMS

Chapter 12

Disturbance Rejection for Magnetic Leviation Suspension System

12.1 Introduction

In the past few years, MAGnetic LEViation (MAGLEV) suspension systems has been attracting more and more attention as a promising transportation system [146]. Compared with conventional trains, the major superiority of MAGLEV train lies in that the friction, mechanical losses, vibration, and noise are reduced substantially since it replaces the wheels by electromagnets, levitates on the guideway and avoids mechanical contact with the rail [147]. However, MAGLEV suspensions are essentially nonlinear systems with lumped disturbances consisting of external disturbances (caused by track input) and parameter perturbations (caused by load variation) [148, 149]. A number of elegant control approaches for MAGLEV systems have been researched throughout the past two decades, including PI control [103], sliding mode control [32], adaptive control [33], robust control [150], H-infinity control [34, 35], and some other traditional methods [36, 37].

Two reasons motivate the investigation of DOBC method to the MAGLEV suspension systems. One is most of the feedback-based control methods listed above can not react promptly in the presence of strong disturbances since they do not deal with the disturbance directly. The second is that the traditional DOBC methods are possibly unavailable due to the reason that the lumped disturbances in MAGLEV systems are mismatched ones. This chapter mainly refers to [12].

12.2 Problem Formulation

12.2.1 Nonlinear MAGLEV Suspension Dynamics

The complete nonlinear model for the MAGLEV suspension system is given by [37, 103],

$$B = K_b \frac{I}{G}, \tag{12.1}$$

$$F = K_f B^2, \tag{12.2}$$

$$\frac{dI}{dt} = \frac{V_c - IR_c + \frac{N_c A_p K_b}{G^2}\left(\frac{dz_t}{dt} - \frac{dZ}{dt}\right)}{\frac{N_c A_p K_b}{G} + L_c}, \tag{12.3}$$

$$\frac{d^2 Z}{d^2 t} = g - \frac{K_f}{M_s}\frac{I^2}{G^2}, \tag{12.4}$$

$$\frac{dG}{dt} = \frac{dz_t}{dt} - \frac{dZ}{dt}, \tag{12.5}$$

where variables I, z_t, Z, $\frac{dz_t}{dt}$, $\frac{dZ}{dt}$, G, B, and F denote the current, the rail position, the electromagnet position, the rail vertical velocity, the electromagnet vertical velocity, the air gap, the flux density, and the force, respectively. Signal V_c is the voltage of the coil. Remaining symbols in Equations (12.1)–(12.5) are system parameters which are listed in Table 12.1. The diagram of the MAGLEV suspension system is shown by Figure 12.1.

Table 12.1 Parameters of MAGLEV Suspension System

Parameters	Meaning	Value
M_s	Carriage mass	1000kg
K_b	Flux coefficient	0.0015T·m/A
K_f	Force coefficient	9810N/T²
R_c	Coil's resistance	10Ω
g	Gravity constant	9.81m/s²
L_c	Coil's inductance	0.1H
N_c	Number of turns	2000
A_p	Pole face area	0.01m²

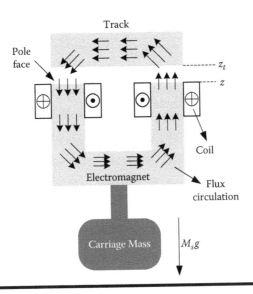

Figure 12.1 Diagram of the MAGLEV suspension system.

12.2.2 Model Linearization

The linearization of the MAGLEV suspension is based on small perturbations around the operating point [103, 149]. The following definitions are used in which the lower case letters define a small variation around the operating point and the subscript "o" refers to the operating condition.

$$B = B_o + b, \tag{12.6}$$

$$F = F_o + f, \tag{12.7}$$

$$I = I_o + i, \tag{12.8}$$

$$G = G_o + (z_t - z), \tag{12.9}$$

$$V_c = V_o + u_c, \tag{12.10}$$

The nominal values of variables in operating point are given in Table 12.2. The nonlinear MAGLEV suspension models (12.3)–(12.5) are expressed as a linear system with disturbances and uncertainties

$$\begin{cases} \dot{x} = Ax + B_u u + B_d d + \Delta Ax + O(x, u, d), \\ y = Cx, \end{cases} \tag{12.11}$$

where the states are the linearized current, vertical carriage velocity, and air gap, i.e., $x = [i \; \dot{z} \; (z_t - z)]^T$, the input $u = u_c$ is the voltage, the track input disturbance

Table 12.2 Nominal Values of MAGLEV Suspension System

Parameters	Meaning	Value
B_o	Nominal flux density	1.0T
F_o	Nominal force	9810 N
I_o	Nominal current	10A
G_o	Nominal air gap	0.015m
V_o	Nominal voltage	100V

$d = \dot{z}_t$ is the rail vertical velocity (the external disturbance). The controlled variable is the variation of the air gap, i.e., $y = z_t - z$. ΔA is the perturbation matrix caused by load variation. Nonlinear function $O(x, u, d)$ consists of the high order nonlinear terms with respect to x, u, and d. The detailed linearization procedure can be found in [103]. The matrixes A, B_u, B_d, and C are given as

$$A = \begin{bmatrix} \frac{-R_c}{L_c + K_b N_c \frac{A_p}{G_o}} & \frac{-K_b N_c A_p I_o}{G_o^2(L_c + K_b N_c \frac{A_p}{G_o})} & 0 \\ -2K_f \frac{I_o}{M_s G_o^2} & 0 & 2K_f \frac{I_o^2}{M_s G_o^3} \\ 0 & -1 & 0 \end{bmatrix}, \qquad (12.12)$$

$$B_u = \begin{bmatrix} \frac{1}{L_c + K_b N_c \frac{A_p}{G_o}} \\ 0 \\ 0 \end{bmatrix}, \qquad (12.13)$$

$$B_d = \begin{bmatrix} \frac{K_b N_c A_p I_o}{G_o^2(L_c + K_b N_c \frac{A_p}{G_o})} \\ 0 \\ 1 \end{bmatrix}, \qquad (12.14)$$

$$C = [0\ 0\ 1]. \qquad (12.15)$$

Remark 12.1 *In (12.11), the high-order nonlinear function $O(x, u, d)$ is generally neglected when designing controllers since their magnitudes are much smaller as compared with the linear parts. In this book, such high-order nonlinear terms $O(x, u, d)$ are no longer neglected but considered as parts of the lumped disturbances.*

The control specifications of the MAGLEV suspension under consideration of deterministic track input are given in Table 12.3.

Table 12.3 Control Specifications for MAGLEV Suspension System (Deterministic)

Constraints	Value
Maximum air gap deviation, $(z_t - z)_p$	≤ 0.0075 m
Maximum input coil voltage, $(u_c)_p$	$\leq 300V(3I_oR_c)$
Settling time, t_s	$\leq 3s$
Air gap steady state error, $(z_t - z)_{ess}$	$= 0$

12.2.3 *Problem Formulation*

In MAGLEV suspension systems, the lumped disturbances (or uncertainties) are formulated by merging the effects caused by parameter variation and nonlinearities into the disturbance terms, which are described as

$$d_l = B_d d + \Delta Ax + O(x, u, d). \tag{12.16}$$

Substituting (12.16) into (12.11), the full dynamic model of the nonlinear MAGLEV suspension system is described as

$$\begin{cases} \dot{x} = Ax + B_u u + B_{ld} d_l, \\ y = Cx, \end{cases} \tag{12.17}$$

where $B_{ld} = I$ is a 3×3 identity matrix.

Remark 12.2 *Note that d_l in (12.17), which is a vector with the dimension three, includes system uncertainties consisting of the external disturbance and the effects caused by parameter variation and nonlinearities.*

Remark 12.3 *It can be observed from Equations (12.13) and (12.17) that there are multiple disturbances but a single control input. Hence, the disturbances definitely enter the system via different channel from the control input, forming a set of mismatched disturbances.*

In the presence of mismatched disturbances/uncertainties, the existing DOBC methods are no longer available. This is better explained via the following example. Considering a simple system, that is,

$$\begin{cases} \dot{x}_1 = x_2 + d, \\ \dot{x}_2 = x_1 + x_2 + u, \\ y = x_1. \end{cases} \tag{12.18}$$

For system (12.18), the estimate $\hat{d}$ of the real disturbance d can be obtained by a disturbance observer. However, if the composite control law is designed as $u = K_x x - \hat{d}$ (where K_x is the feedback control gain) which is the formulation of the existing DOBC methods from the literature, it can be found that the disturbance compensation design has nothing meaningful for system (12.18) because the disturbance can neither be attenuated from the state equations nor from the output channels.

It should be pointed out that the mismatched disturbances are unlikely to be attenuated from the state equations in general. This work aims to develop a disturbance compensation gain so that the DOBC method can be used to remove mismatched disturbances/uncertainties from the output channels.

12.3 DOBC Design

Consider the MAGLEV system with multiple disturbances, depicted by (12.17).

Remark 12.4 *In (12.17), the lumped disturbances are generalized concepts of disturbances, possibly including external disturbances, unmolded high-order dynamics, parameter variations, and nonlinear dynamics that can not be captured by the linear dynamics in Equation (12.17).*

For system (12.17), the following state-space disturbance observer [2, 5, 40, 10] is designed to estimate the disturbances

$$
\begin{cases}
\dot{p} = -L B_{ld}(p + Lx) - L(Ax + B_u u), \\
\hat{d}_l = p + Lx,
\end{cases} \tag{12.19}
$$

where $\hat{d}_l$ is the disturbance estimate vector, p is an auxiliary vector and L is the observer gain matrix to be designed.

Remark 12.5 *The state-space DO presented in (12.19) can be used to estimate both matched and mismatched uncertainties.*

Based on the disturbance estimate by DO, a new composite control law is proposed, that is,

$$
u = K_x x + K_d \hat{d}_l, \tag{12.20}
$$

An appropriately designed disturbance compensation gain K_d guarantees that the influence of the disturbances can be removed from the output channels in the steady-state.

For system (12.17) subject to mismatched uncertainties, a general design procedure of the proposed DOBC method is given below:

1. Design a feedback controller to achieve stability without considering the lumped disturbances.
2. Design a linear state-space disturbance observer to estimate the lumped disturbances.
3. Design a disturbance compensation gain to guarantee that the mismatched lumped disturbances are removed from the output channels in steady-state.
4. Integrate the feedback controller and the feedforward compensator to formulate the composite DOBC law.

To establish the stability of the closed-loop system, some mild assumptions are given as follows.

Assumption 12.1 *Both the lumped disturbance d_l and its derivative $\dot{d}_l$ are bounded.*

Assumption 12.2 *The lumped disturbances d_l have constant values in steady-state, i.e.,* $\lim\limits_{t\to\infty} \dot{d}_l(t) = 0$ *or* $\lim\limits_{t\to\infty} d_l(t) = d_s$ *where d_s is a constant vector.*

Assumption 12.3 (A, B_u) *is controllable.*

The asymptotic stability of DO (12.19) is concluded by the following theorem.

Theorem 12.1 *Suppose that Assumptions 12.1 and 12.2 are satisfied for system (12.17). The disturbance estimates $\hat{d}_l$ yielded by DO (12.19) can asymptotically track the lumped disturbances d_l if the observer gain matrix L in (12.19) is chosen such that matrix $-LB_{ld}$ is Hurwitz.*

Proof The disturbance estimation error of the DO (12.19) is defined as

$$e_d = \hat{d}_l - d_l. \tag{12.21}$$

Combining system (12.17), DO (12.19), with estimation error (12.21), gives

$$
\begin{aligned}
\dot{e}_d &= \dot{\hat{d}}_l - \dot{d}_l \\
&= \dot{p} + L\dot{x} - \dot{d}_l \\
&= -LB_{ld}\hat{d}_l - L(Ax + B_u u) + L(Ax + B_u u + B_{ld}d_l) - \dot{d}_l \\
&= -LB_{ld}(\hat{d}_l - d_l) - \dot{d}_l \\
&= -LB_{ld}e_d - \dot{d}_l.
\end{aligned} \tag{12.22}
$$

It can be verified that the error system (12.22) is asymptotically stable since $-LB_{ld}$ is Hurwitz, $\dot{d}_l$ is bounded and satisfies $\lim\limits_{t\to\infty} \dot{d}_l = 0$. This implies that the disturbance estimate of DO can track the disturbances asymptotically.

Remark 12.6 *It seems that the asymptotic tracking of DO can only be obtained based on the assumption that* $\lim\limits_{t\to\infty} \dot{d}_l(t) = 0$. *However, it is pointed out in [2] that the* *estimates by DO (12.19) can also track some fast time-varying disturbances well as long as the observer dynamics are much faster than those of the disturbances.*

Remark 12.7 *In the presence of uncertainties, the lumped disturbances would be a function of the states, which can be reasonably estimated if the DO dynamics are faster than the closed-loop dynamics. The same argument for the state observer based control methods is applicable.*

The bounded-input-bounded-output (BIBO) and asymptotic stabilities of the closed-loop system are shown as follows.

Theorem 12.2 *Suppose that Assumptions 12.1 and 12.3 are satisfied for system (12.17). The BIBO stability of system (12.17) under the newly proposed DOBC law (12.20) is guaranteed if the observer gain L in (12.19) and the feedback control gain K_x in (12.20) are selected such that both $-LB_{ld}$ and $A + B_u K_x$ are Hurwitz.*

Proof Combining system (12.17), composite control law (12.20), with error system (12.22), the closed-loop system is written as

$$\begin{bmatrix} \dot{x} \\ \dot{e}_d \end{bmatrix} = \begin{bmatrix} A + B_u K_x & B_u K_d \\ 0 & -LB_{ld} \end{bmatrix} \begin{bmatrix} x \\ e_d \end{bmatrix} + \begin{bmatrix} B_u K_d + B_{ld} & 0 \\ 0 & -1 \end{bmatrix} \begin{bmatrix} d_l \\ \dot{d}_l \end{bmatrix}. \quad (12.23)$$

Since $-LB_{ld}$ and $A + B_u K_x$ are Hurwitz, it can be verified that matrix

$$\begin{bmatrix} A + B_u K_x & B_u K_d \\ 0 & -LB_{ld} \end{bmatrix}$$

is also Hurwitz. It can be concluded that the closed-loop system (12.23) is BIBO stable for any bounded d_l and $\dot{d}_l$ if K_x and L are properly selected.

Theorem 12.3 *Suppose that Assumptions 12.1 and 12.2 are satisfied for system (12.17). The states of system (12.17) under the composite control law (12.20) converge to $x_s = -(A + B_u K_x)^{-1}(B_u K_d + B_{ld})d_s$ asymptotically if the observer gain L in (12.19) and the feedback control gain K_x in (12.20) are selected such that both $-LB_{ld}$ and $A + B_u K_x$ are Hurwitz.*

Proof The state error can be constructed as

$$e_x = x - x_s. \quad (12.24)$$

Combining system (12.17), composite control law (12.20), error system (12.22), with state error (12.24), the closed-loop system is given as

$$\begin{bmatrix} \dot{e}_x \\ \dot{e}_d \end{bmatrix} = \begin{bmatrix} A + B_u K_x & B_u K_d \\ 0 & -LB_{ld} \end{bmatrix} \begin{bmatrix} e_x \\ e_d \end{bmatrix} + \begin{bmatrix} B_u K_d + B_{ld} & 0 \\ 0 & -1 \end{bmatrix} \begin{bmatrix} d_l - d_s \\ \dot{d}_l \end{bmatrix}. $$
$$(12.25)$$

Similar with the proof of Theorem 12.2, it can be verified that

$$
\begin{bmatrix}
A + B_u K_x & B_u K_d \\
0 & -L B_{ld}
\end{bmatrix}
$$

is Hurwitz. In addition, it can be obtained from Assumption 12.2 that $\lim\limits_{t\to\infty} [d_l(t) - d_s] = 0$.

Considering the conditions, it can be shown that the closed-loop system (12.25) is asymptotically stable. This means that with the given conditions the state vector converges to a constant vector x_s asymptotically.

The main theoretical contribution of this work is that a systematic method is developed for disturbance compensation gain design so that the disturbances can be attenuated from the output channels in steady-state.

Assumption 12.4 *The system matrices and the feedback control gain satisfy the rank condition that* $rank(C(A + B_u K_x)^{-1} B_u) = rank([C(A + B_u K_x)^{-1} B_u, -C(A + B_u K_x)^{-1} B_{ld}])$.

Theorem 12.4 *Suppose that Assumptions 12.1–12.4 are satisfied for system (12.17), also the observer gain L and the feedback control gain K_x are chosen to make matrixes $-L B_{ld}$ and $A + B_u K_x$ Hurwitz. Considering system (12.17) under the proposed composite control law (12.20), the lumped disturbances can be attenuated from the output channel in steady-state if the disturbance compensation gain is designed such that*

$$
C(A + B_u K_x)^{-1} B_u K_d = -C(A + B_u K_x)^{-1} B_{ld}. \tag{12.26}
$$

Proof Substituting control law (12.20) into system (12.17), the state is expressed as

$$
x = (A + B_u K_x)^{-1} \left[\dot{x} - B_u K_d \hat{d}_l - B_{ld} d_l \right]. \tag{12.27}
$$

Combining (12.17), (12.26) with (12.27), gives

$$
y = C(A + B_u K_x)^{-1} \dot{x} + C(A + B_u K_x)^{-1} B_{ld}(\hat{d}_l - d_l). \tag{12.28}
$$

With the given conditions, the following results are obtained according to Theorem 12.3,

$$
\lim_{t\to\infty} e_x(t) = 0, \ \lim_{t\to\infty} e_d(t) = 0. \tag{12.29}
$$

Under the consideration of (12.29), taking limits of both sides of Equation (12.28) with respect to t, gives

$$
\lim_{t\to\infty} y(t) = 0. \tag{12.30}
$$

Remark 12.8 *The disturbance compensation gain K_d in (12.26) is a general case and suitable for both matched and mismatched disturbances. In the matching case, that is, $B_u = B_{ld}$, it can be obtained from (12.26) that the disturbance compensation gain is reduced to $K_d = -1$ which is the particular form of most existing DOBC methods.*

It should be pointed out that any feedback controller that can stabilize system (12.17) in the absence of disturbances may be used to achieve tracking performance. Here we concentrate on deterministic design only. In this study, the classical linear quadratic regulator (LQR) is employed and the penalty matrixes Q and R in the cost function of LQR are simply selected as

$$
Q = \begin{bmatrix} 1 & 0 & 0 \\ 0 & 1 & 0 \\ 0 & 0 & 1 \end{bmatrix}, \quad R = 0.1. \tag{12.31}
$$

To this end, the feedback control gain of LQR can be calculated according to the model information and the penalty matrixes, given as $K_x = [-61 \ -591 \ 40061]$. With the parameter matrixes given in (12.12), (12.13), (12.17) and the calculated control gain K_x, the disturbance compensation gain is calculated by Equation (12.26), given as $K_d = [-2.1 \ 36.0 \ 742.2]$.

To guarantee the asymptotic stability of the DO, the observer gain matrix in (12.19) is designed as

$$
L = \begin{bmatrix} 40 & 0 & 0 \\ 0 & 40 & 0 \\ 0 & 0 & 40 \end{bmatrix}. \tag{12.32}
$$

The control structure of the proposed DOBC method for the nonlinear MAGLEV suspension system is shown in Figure 12.2.

12.4 Simulations and Analysis

In the presence of disturbances/uncertainties, many control methods (e.g., LQR) can not remove the offset. The integral control is a widely used practical method to eliminate the steady-state error in these cases. To demonstrate the effectiveness of the proposed method, LQR plus an integral action (called LQR+I) method is employed for comparison. The control law of the LQR+I method is represented as

$$
u_{lqr+i}(t) = K_x x(t) + k_i \int_{t_0}^{t} [z_t(\tau) - z(\tau)] \, d\tau, \tag{12.33}
$$

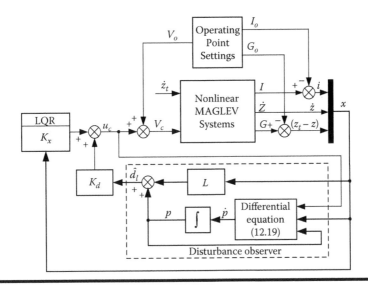

Figure 12.2   Block diagram of the proposed DOBC method for the nonlinear MA-GLEV system.

where t_0 is the initial time and k_i is the integral coefficient to be designed. To find an appropriate integral coefficient, a variable $x_4 = \int_{t_0}^{t} [z_t(\tau) - z(\tau)]\, d\tau$ is taken as an augmented state. For the augmented system, the LQR method is used to get the control gain. The penalty matrixes are selected as

$$Q_I = \begin{bmatrix} Q & 0 \\ 0 & q_i \end{bmatrix}, \quad R_I = R = 0.1. \tag{12.34}$$

Set $q_i = 1.0 \times 10^9$, then the integral coefficient is able to be calculated, which is given as $k_i = -1.0 \times 10^5$.

12.4.1 *External Disturbance Rejection Performance*

The main external disturbances in the MAGLEV system are the deterministic inputs to the suspension in the vertical direction. Such deterministic inputs are the transitions onto track gradients. Here, the deterministic input components considered are referred to [103] and shown in Figure 12.3. They represent a gradient of 5% at a vehicle speed of 15 *m/s* and an allowed acceleration of 0.5 *m/s*² while the jerk level is 1 *m/s*³.

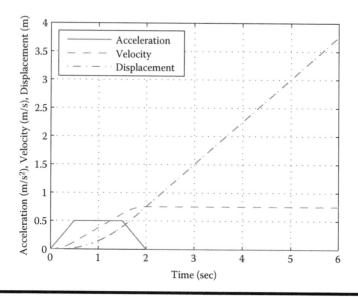

Figure 12.3 **Track (deterministic) input to the suspension with a vehicle speed of 15 m/s and 5% gradient.**

The response curves of both the output and input of the suspension system under the proposed DOBC method (solid line), LQR+I method (dashed line) and LQR method (dash-dotted line) are shown in Figure 12.4. Response curves of the corresponding states are shown in Figure 12.5. Note that zero air gap variation refers to G_o, as we follow the treatment setup proposed in [37, 148, 149].

As shown in Figures 12.4 and 12.5, the LQR method results in unstable control of the airgap and current in such case of external disturbances. It can be observed from Figure 12.4(a) that both the overshoot and settling time under the DOBC method are shorter than those under the LQR+I method, the convergence rate under the DOBC method is much faster than that under the LQR+I method. As shown in Figure 12.4(b), the maximum input voltage under the DOBC method is smaller than that under the LQR+I method. Response curves in Figure 12.5 show that both the current and the vertical electromagnet velocity under the DOBC method vary smoothly and approach to the desired equilibrium points faster than that under the LQR+I method. The results demonstrate that the newly proposed DOBC method has achieved a much better performance in rejecting such practical disturbances than that of the traditional LQR+I method.

12.4.2 Robustness Against Load Variation

In this part, the load variation of the MAGLEV suspension is considered. The suspension has to support the large mass of the vehicle as well as the load (weight of

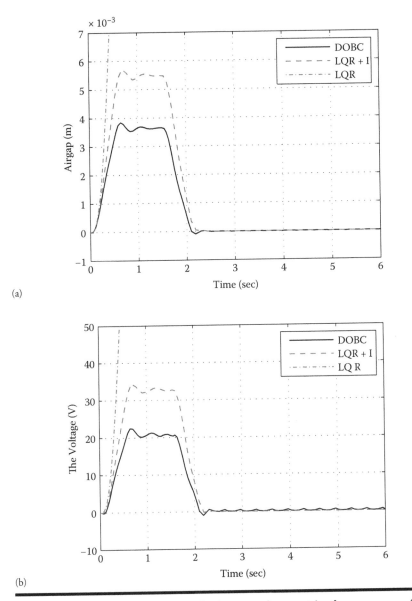

(a)

(b)

Figure 12.4 Response curves of the input and output in the presence of deter-ministic track input: (a) the air gap, $(z_t - z)$, (b) the voltage of the coil , u_c.

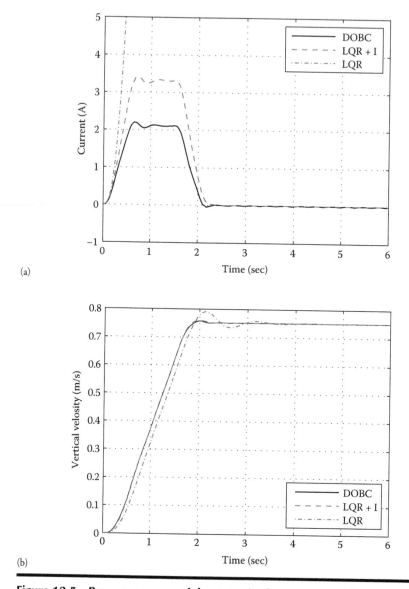

Figure 12.5 **Response curves of the states in the presence of deterministic track input: (a) the current, *i*, (b) the vertical electromagnet velocity, *ż*.**

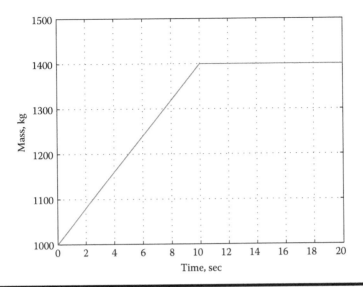

Figure 12.6   Curve of the load variation.

the passengers) which can vary up to 40% of the total mass of the vehicle [103]. This is a considerable variation of the total mass and may result in undesirable performance. To this end, the robustness against load variations should be taken into account to ensure performance and stability for a fully laden or unladen vehicle. For testing, it is supposed that the load variation is up to 40% of the total vehicle mass within 10 seconds, that is, the load can vary from $1000kg$ to $1400kg$ for a fully unladen and laden vehicle, respectively. The profile of the load variation is shown in Figure 12.6. The track input disturbance like in Figure 12.3 is also considered but acts on the system at $t = 15$ s.

The robustness against such case of load variation under both DOBC (solid line), LQR+I (dashed line) and LQR (dash-dotted line) methods can be seen in Figures 12.7 and 12.8. It can be found that the LQR method leads to unstable control of the closed-loop system in such cases of uncertainties. It can be observed from Figure 12.7(a) that the overshoot of the air gap deviation under DOBC method is much smaller than that under the LQR+I method. In addition, the convergence rate under the DOBC method is much faster than that under the LQR+I method. Figure 12.7(b) shows that the magnitude of the coil voltage under DOBC method is smaller than that under the LQR+I method. Test results in this subsection manifest that the proposed method obtains promising performance of robustness against load variation as compared with that of the integral control method.

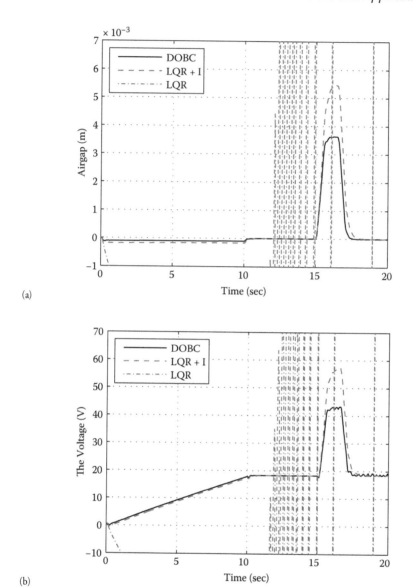

(a)

(b)

Figure 12.7 **Response curves of the input and output in the presence of load variation: (a) the air gap, $(z_t - z)$, (b) the voltage of the coil , u_c.**

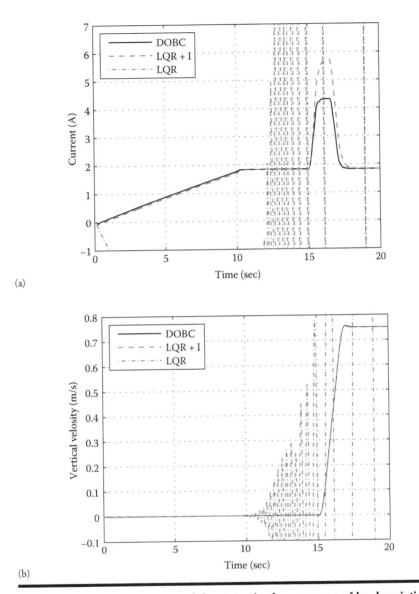

(a)

(b)

Figure 12.8 **Response curves of the states in the presence of load variation: (a) the current, *i*, (b) the vertical electromagnet velocity, *ż*.**

12.5 Summary

By appropriately designing a disturbance compensation gain, a novel disturbance observer-based control (DOBC) method has been proposed to solve the disturbance, rejection problem of the uncertain system that contains mismatched disturbances. Here, mismatched means the lumped disturbances enter the system through different channels from the control inputs. With the proposed control method, rigorous stability of the closed-loop system has been analyzed under some soundable assumptions. It has also been verified that the mismatched lumped disturbances can be attenuated from the output channels with a properly designed disturbance compensation gain. To shown the feasibility and efficiency of the proposed method, application design is carried out for an industrial MAGLEV suspension system, which is essentially a system with mismatched disturbances including external disturbances (caused by track inputs), nonlinear dynamics (neglected nonlinearities during liberalization), and parameter perturbations (caused by load variation). When controlled by such practical MAGLEV suspension system, simulations have shown that the proposed method achieves a much better performance of external disturbance rejection and robustness against load variation than those of the traditional integral control method.

Chapter 13

Disturbance Rejection for Permanent Magnet Synchronous Motors

13.1 Introduction

A permanent magnet-synchronous motor has many excellent features such as a simple structure, high torque-to-current radio, high power density, low noise, and friendly maintenance [151]. Due to these advantages, PMSM has gained widespread acceptance in numerical control machine tools, robots, aviation, and so on. Linear control methods such as the proportional-integral (PI) control scheme are already widely used in PMSM systems due to their easy implementation. However, since PMSM system is a nonlinear system with unavoidable and unmeasured disturbances and parameter variations [152], it is difficult to achieve a satisfactory performance in the entire operating rage when using such linear control methods [153, 154]. Hence, nonlinear control methods become natural improved solutions for the PMSM system.

With the development of microprocessor technology, especially digital signal processors (DSP), power electronics, and modern control theories, more and more advanced control methods are introduced to the PMSM control problem, e.g., adaptive control [155, 156, 157, 158], robust control [159], sliding mode control [160, 161], input-output linearization control [153, 162], backstepping control [163, 164], fractional order control [165], neural network control [154], fuzzy control [166], and finite-time control [167], etc. These methods can improve the performance of the PMSM systems from different aspects.

Note that in real industrial applications, PMSM systems always face with different disturbances, e.g., friction force, unmodeled dynamics, and load disturbances.

Compared with conventional feedback-based control methods, which usually can not react promptly to reject these disturbances, control methods based on feedforward compensation techniques for disturbances show a direct, fast, and effective way to suppress the unfavorable influences caused by disturbances. Under this framework, many disturbance rejection control methods have been reported.

Two kinds of disturbance estimation techniques have been well developed to facilitate the design of such composite control methods since in real PMSM applications it is usually impossible to measure the disturbances. One is the disturbance observer (DO) technique originally presented by Ohnishi in 1987 [168]. Following this direction, many DO-based control methods have been reported in different applications, e.g., robotic systems [169], hard disk drive systems [170], general motion control systems [5, 85, 171, 172, 97], inverted pendulum systems [87], PMSM systems [173, 152, 174, 157], etc. Among these results, different kinds of DO have been developed, including linear DOs [173, 152, 171, 170], fuzzy DOs [87], neural network-based DOs [174], and other nonlinear DOs [169, 5, 85, 172, 157, 97], etc.

Another kind of disturbance estimation technique is the extended state observer (ESO) originally presented by Han in 1995 [9, 4]. It regards the lumped disturbances of a system, which consists of internal dynamics and external disturbances, as a new state of system. It can estimate both the states and the lumped disturbances. A composite control frame based on ESO, called active disturbance rejection control (ADRC), is also developed. This method has also been applied in many industrial control problems, e.g., robotic systems [175], machining processes [176], manipulator systems [177], power converters [178], PMSM systems [156, 161, 167, 179], other motor systems [180, 181], and chemical systems [182].

There are quite a few different usages for disturbance estimation techniques in the control design of PMSM systems. In [179], the active disturbance rejection controller is designed for the position loop of PMSM system. This controller consists of three parts: a tracking differentiator, a feedforward part based on a nonlinear ESO, and a feedback part based on nonlinear proportional derivative control. For the precise position control problem of PMSM system, a torque disturbance observer based on neural network is constructed to produce a feedforward part for the position control to improve the ability against load torque and parameter variations. In [156], for the speed regulation control of PMSM system, an ESO-based composite controller is designed for the speed loop. The speed controller consists of a feedforward part based on a linear ESO and a feedback part based on linear proportional control. Furthermore, considering the variations of inertia, an adaptive control scheme is developed by using inertia identification techniques, the gain of feedforward compensation based on the ESO can be auto-tuned according to the identified inertia. In [167], a composite speed controller consisting of a feedforward part based on a linear ESO and a feedback part based on finite-time proportional control is developed for the speed regulation problem of PMSM system. In [161], considering a second-order model of permanent magnet synchronous motor system, a composite speed controller consisting of a sliding mode feedback part and a feedforward compensation part based on a linear ESO is proposed. In [152], two linear disturbance observers are constructed to estimate load torque and flux linkage, respectively. These two estimate variables are employed in the feedback linearization control design part, which helps to compensate the nonlinear disturbances caused by the incomplete linearization. A nonlinear speed control law is thus developed.

While most of the existing researches focus on introducing advanced control methods to the position loop or speed loop, fewer attention has been paid to the current loops. Note that the current loops also have parameters variations, e.g., variations of stator resistance, inaccurate back-electromotive force (EMF) model. These parameter variations may primarily degrade the control performance [157]. Linear control algorithms like PI control are widely employed because of simple calculation and easy implementation. Although the real time requirements may restrict the application of advanced control method to current loops, fortunately, owing to the fast development of computing hardware, it becomes possible to implement these controllers in current loops. There are reports on the application of different advanced control algorithms in current controls for PMSM systems, including model predictive control [183, 184], model reference adaptive control [158], with simulation and experimental verification results.

Almost all of the disturbance estimation results are for the position loop or speed loop, few results are for the current loop except [157, 173]. In [173], a linear disturbance observer based on state equation is developed for estimating the disturbances of the two current loops. Thus, two composite control laws are obtained for the two current loops, respectively. In [157], a composite control law combining PI feedback part and feedforward compensation part based on adaptive disturbance estimation is developed for q-axis current loop. Any improvement on these inner loops in rejecting parameter variations will effectively help the performance improvement of the whole system.

In this chapter, to pursue a high disturbance rejection ability, a control scheme which employs disturbance-rejection control laws for not only speed loop but also the q-axis current loop is proposed. For the speed loop, regarding parameter variations and load torque as the lumped disturbances, an ESO-based disturbance rejection control law is developed. Considering the dynamics of q-axis current, the coupling between rotor speed and d-axis current as well as the back electromotive force are regarded as lumped disturbances for the q-axis current loop, lumped disturbances are estimated by introducing an ESO. Thus, a composite control law consisting of proportional feedback and disturbance feedback compensation simplified model is developed to control the q-axis current. Simulation and experiment comparisons are presented to verify the effectiveness of the proposed method.

13.2 Problem Description

Assume that magnetic circuit is unsaturated, hysteresis and eddy current loss are ignored and the distribution of the magnetic field is sine space. Under this condition, in d-q coordinates, the ideal model of the surface-mounted permanent magnet synchronous motor is expressed as follows [185]

$$
\begin{pmatrix} \dot{i}_d \\ \dot{i}_q \\ \dot{\omega} \end{pmatrix} = \begin{pmatrix} -\frac{R_s}{L_d} & n_p\omega & 0 \\ -n_p\omega & -\frac{R_s}{L_q} & -\frac{n_p\psi_f}{L_q} \\ 0 & \frac{3n_p\psi_f}{2J} & -\frac{B}{J} \end{pmatrix} \times \begin{pmatrix} i_d \\ i_q \\ \omega \end{pmatrix} + \begin{pmatrix} \frac{u_d}{L_d} \\ \frac{u_q}{L_q} \\ -\frac{T_L}{J} \end{pmatrix}
\tag{13.1}
$$

where

R_s = the stator resistance;
u_d, u_q = the d- and q- axes stator voltages;
i_d, i_q = the d- and q- axes stator currents;
L_d, L_q = the d- and q- axes stator inductances, and $L_d = L_q = L$;
n_p = the number of pole pairs of the PMSM;
ω = the rotor angular velocity of the motor;
ψ_f = the flux linkage;
T_L = the load torque;
B = the viscous friction coefficient;
J = the rotor inertia.

The general structure of the PMSM servo system is shown in Figure 13.1. The overall system consists of a PMSM with load, space vector pulse-width modulation (SVPWM), voltage-source inverter (VSI), field-orientation mechanism, and three controllers. The controllers employ a cascade control loop structure, including a speed loop and two current loops. Here, the PI controller, which is used to stabilize the d-axis current errors of the vector controlled drive, is adopted in the d axis current loop. As it can be seen from Figure 13.1, the rotor angular velocity ω can

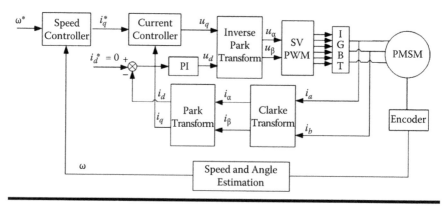

Figure 13.1 The principle block diagram of PMSM system based on vector control.

be obtained from the position and speed sensor. The currents i_d and i_q can be calculated from i_a and i_b (which can be obtained from measurements) by Clarke and Park transformations. Here, $i_d^* = 0$ strategy is used, and we mainly design controllers for the speed loop and the q-axis current loop.

13.3 Control Strategy

There exist different kinds of disturbances in real applications of PMSM systems, including internal disturbances, such as parameters variation, back electromotive forces, torque ripples, friction forces, and unmodeled dynamics, and external disturbances, like load disturbances. These disturbances may degrade the performance of a closed-loop system if the controller does not have enough ability to reject them. Since the control structure employed here is of cascade control loops for the speed regulation problem, the control scheme includes a speed loop and two current loops. For the d-axis current loop, PI control algorithm is employed. For the speed loop and q-axis current loop, two ESOs are introduced to estimate the different disturbances in each loop, respectively; furthermore two composite control law are developed.

13.3.1 The Principle of ESO-Based Control Method

The main disturbance estimation technique here is ESO. It concerns the lumped disturbances of system, which consists of internal dynamics and external disturbances, as a new state of system. This observer is one order more than the usual state observer. It can estimate both the states and the lumped disturbances. Based on the ESO, a feed-forward compensation for the disturbances can also be employed in the control design.

Taking the first-order system, for example, how to construct linear ESO will be shown below. For the following first-order system

$$\dot{x} = f(x, t) + d(t) + bu(t) \tag{13.2}$$

where $f(x, t)$ is the unknown nonlinear function, $d(t)$ is the external disturbance, u is the control signal, x is the output of system, and b is the control gain. Considering the lumped disturbance $a(t) = f(x, t) + d(t)$ as an extended state, then Equation (13.2) can be expressed as

$$\dot{x} = a(t) + bu(t). \tag{13.3}$$

Define $x_1 = x$, $x_2 = a(t)$. The system (13.3) can be rearranged as

$$\begin{cases} \dot{x}_1 = x_2 + bu(t) \\ \dot{x}_2 = c(t) \end{cases} \tag{13.4}$$

where $c(t) = \dot{a}(t)$. A linear ESO can be constructed as follows [4]:

$$\begin{cases} \dot{z}_1 = z_2 - 2p(z_1 - x) + bu \\ \dot{z}_2 = -p^2(z_1 - x) \end{cases} \tag{13.5}$$

where $-p$ ($p > 0$) is the desired double pole of ESO, z_1 is the estimate of x_1, and z_2 is the estimate of $a(t)$. Here, according to the analysis in [186], $z_1(t)$ is an estimate of the output $x(t)$, while $z_2(t)$ is an estimate of the lumped disturbance $a(t)$, i.e., $z_1(t) \to y(t)$ and $z_2(t) \to a(t)$.

A composite control law is designed as follows

$$u = u_0 - \frac{z_2}{b}, \quad u_0 = k(x^* - z_1) \tag{13.6}$$

where x^* is the reference input of x.

13.3.2 Speed Controller Design

The main disturbances in the speed loop include parameter variations, uncertainties, friction, tracking error from current-loops, and load torque variations. In order to improve the performance of PMSM system, an ESO-based controller for the speed loop is designed as follows.

The speed equation of PMSM system is

$$\dot{\omega} = \frac{3n_p \psi_f}{2J} i_q - \frac{B}{J} \omega - \frac{T_L}{J} = bi_q^* + a(t) \tag{13.7}$$

where $b = \frac{3n_p \psi_f}{2J}$ is the control gain, $a(t) = \frac{3n_p \psi_f}{2J}(i_q - i_q^*) - \frac{B}{J}\omega - \frac{T_L}{J}$ is the lumped disturbances consisting of the friction, the external load disturbances, and the tracking error of current-loop of i_q.

Then, an ESO can be designed as

$$\begin{cases} \dot{z}_1 = z_2 - 2p(z_1 - \omega) + bu \\ \dot{z}_2 = -p^2(z_1 - \omega) \end{cases} \tag{13.8}$$

where z_1 is the estimate of angular velocity ω, and z_2 is the estimate of the lumped disturbances.

A composite control law of speed loop can be designed as

$$i_q^* = u = u_0 - \frac{z_2}{b}, \quad u_0 = k(\omega^* - \omega) \tag{13.9}$$

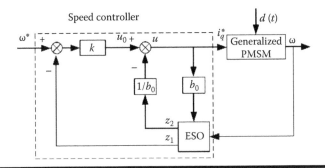

Figure 13.2 **The block diagram of ESO-based speed controller.**

where k is the proportional gain, ω^* is the reference speed, i_q^* is the output of speed loop controller. The block diagram of the ESO-based speed controller is shown in Figure 13.2. Note that the "generalized PMSM" represents the two current loops, which include PMSM and other components the same as that in Figure 13.1.

13.3.3 Current Controller Design

It is well known that the current loops have parameters variations, e.g., variations of stator resistance, inaccurate back-electromotive force (EMF) model. These parameter variations may primarily degrade the control performance [157]. Since the torque (q-axis current) loop is controlled with a response time much shorter than that of the speed loop, and it is an inner loop of the cascade control structure, any improvement on this inner loop in rejecting parameter variations will definitely help the performance improvement of the whole system. In order to suppress unfavorable effects caused by these factors, an ESO-based controller is designed for the q-axis current loop. For the d-axis current loop, a standard PI control algorithm is employed.

The equation of q-axis current loop is written as follows

$$\dot{i}_q = -n_p\omega i_d - \frac{R_s}{L}i_q - \frac{n_p\psi_f}{L}\omega + \frac{u_q}{L} = a_q(t) + \frac{u_q}{L} \qquad (13.10)$$

where the lumped disturbances of q-axis current loop are $a_q(t) = -n_p\omega i_d - \frac{R_s}{L}i_q - \frac{n_p\psi_f}{L}\omega$, consisting of the coupling between rotor speed and d-axis current, and the dynamics of q-axis current, and the back electromotive force.

The ESO of q-axis current loop can be designed as follows

$$\begin{cases} \dot{z}_{11} = z_{12} - 2p_q(z_{11} - i_q) + \frac{u_q}{L} \\ \dot{z}_{12} = -p_q^2(z_{11} - i_q) \end{cases} \qquad (13.11)$$

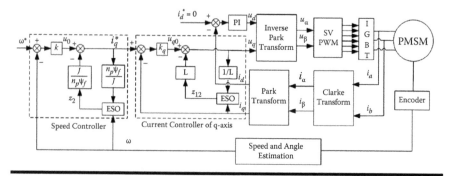

Figure 13.3 The principle diagram of PMSM system under ESO based composite control methods.

where $-p_q$ $(p_q > 0)$ is the desired double pole, z_{11} is the estimate of i_q, and z_{12} is the estimate of $a_q(t)$.

A composite control law of q-axis current loop can be designed as

$$u_q = u_{q0} - Lz_{12}, \quad u_{q0} = k_q(i_q^* - i_q) \tag{13.12}$$

where k_q is the proportional gain of q-axis current loop controller and i_q^* is the q-axis reference current.

Now, the whole speed regulation control scheme of PMSM system is shown in Figure 13.3.

13.4 Simulation and Experimental Results

In order to verify the proposed algorithm, some simulations and experimental studies are done. The results are compared with the control scheme, named as speed ESO based control (speed ESOC) method, which only employs ESO based control law in the speed loop, and use PI algorithms in the both current loops. In order to simplify the description, the proposed method here is named as current ESOC. To have a fair comparison, the control inputs of two schemes have the same saturation limit. Both in simulation and experimental verifications, the parameters of the PMSM are taken the same as follows: the resistance of stator $R_s = 1.74\,\Omega$, the inductances of d and q axes $L_d = L_q = 4\,mH$, the flux of rotor $\psi_f = 0.402\,Wb$, the rotor inertia $J = 1.78 \times 10^{-4}\,kg \cdot m^2$, the viscous friction coefficient $B = 7.4 \times 10^{-5}\,N \cdot m \cdot s$, number of poles $n_p = 4$.

13.4.1 Simulation Results

The speed regulation system of the PMSM vector control is simulated by MATLAB. The saturation limit of q-axis reference current is $11.78\,A$. The proportional and integral gains of d-axis current controllers of the two control schemes are $k_p = 50$, $k_i = 2500$. The parameters of q-axis current loop of the speed ESOC scheme are the same as that of the d-axis current loop. The proportional gain and the desired double-pole of ESO of the speed loop of the speed ESOC scheme are $k = 1.5$ and $p = 5000$. For the current ESOC scheme, the proportional gain and the desired double-pole of the current and speed loops are takes as $k_q = 50$, $p_q = 2000$, and $k = 1.5$, $p = 5000$, respectively. The load torque $T_L = 2.5\,Nm$ is added at $t = 0.1\,s$.

The tracking responses of i_q to i_q^* for the q-axis current loop under speed ESOC and current ESOC methods, respectively, are shown in Figure 13.4. It can be seen that for the current ESOC control method, it takes a shorter time for i_q to track i_q^*. The speed responses of the PMSM system are shown in Figure 13.5. It can be seen that the system using current ESOC has a shorter settling time and a less overshoot. In the presence of load disturbances, for the current ESOC method, the system state can recover much faster with a less speed fluctuations against the disturbances. Figure 13.6 shows the estimation effects of ESO for speed and current loops, respectively.

13.4.2 Experimental Results

To evaluate the performance of the proposed method, the experimental setup system for the speed control of PMSM has been built. The configuration of it and the experimental test setup are shown in Figure 13.7. The whole speed control algorithms including the SVPWM, which are implemented by the program of the DSP TMS320F2808 with a clock frequency of 100MHZ. The control algorithm is implemented using C-program. The PM synchronous motor is driven by a three-phase PWM inverter with an intelligent power module (IPM) with a switching frequency of 10 kHz. The phase currents are measured by the Hall-effect devices and are converted through two 12-bit A/D converters. An incremental position encoder of 2500 lines is used to measure the rotor speed and absolute rotor position.

The saturation limit of q-axis reference current is $11.78\,A$. The proportional and integral gains of d-axis current controllers of the two control schemes are $k_p = 42$ and $k_i = 2600$. The parameters of q-axis current loop of the speed ESOC scheme are the same as that of the d-axis current loop. The proportional gain and the desired double-pole of ESO of the speed loop of the speed ESOC scheme are $k = 1.2$ $p = 330$. For the current ESOC scheme, the proportional gain and the desired

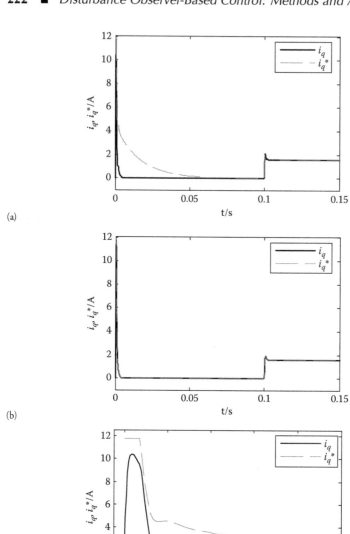

(a)

(b)

(c)

Figure 13.4 Curves of i_q, i_q^* under speed ESOC and current ESOC methods. (a) The whole curves under speed ESOC method. (b) The whole curves under current ESOC method. (c) The local curves of (a) in time domain [0,0.01] s. (d) The local curves of (b) in time domain [0,0.004] s. (e) The local curves of (a) in time domain [0.099,0.103] s. (f) The local curves of (b) in time domain [0.099,0.103] s. (*Continued*)

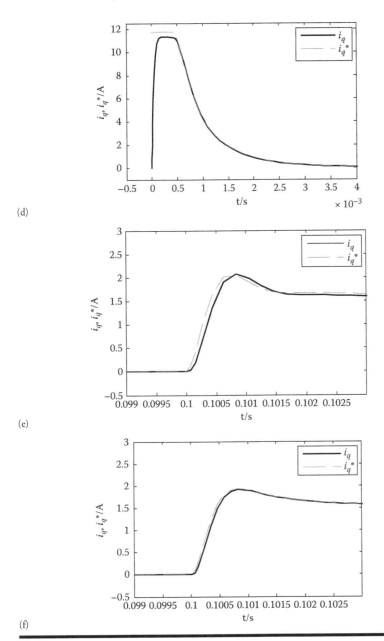

(d)

(e)

(f)

Figure 13.4 (*Continued*) **Curves of i_q, i_q^* under speed ESOC and current ESOC methods. (a) The whole curves under speed ESOC method. (b) The whole curves under current ESOC method. (c) The local curves of (a) in time domain [0,0.01] s. (d) The local curves of (b) in time domain [0,0.004] s. (e) The local curves of (a) in time domain [0.099,0.103] s. (f) The local curves of (b) in time domain [0.099,0.103] s.**

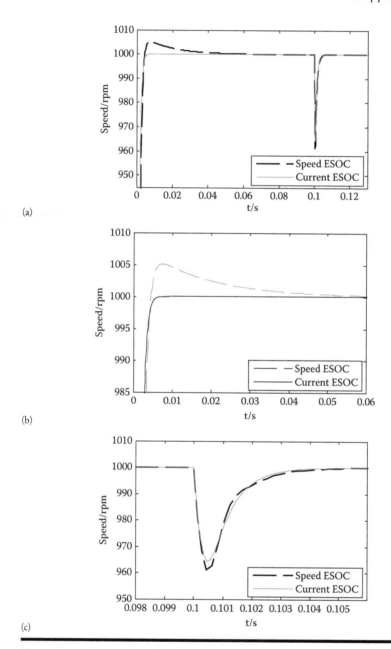

(a)

(b)

(c)

Figure 13.5 Response of speed under speed ESOC and current ESOC methods. (a) the whole curves. (b) the local curves in time domain [0,0.06] s. (c) the local curves in time domain [0.098,0.105] s.

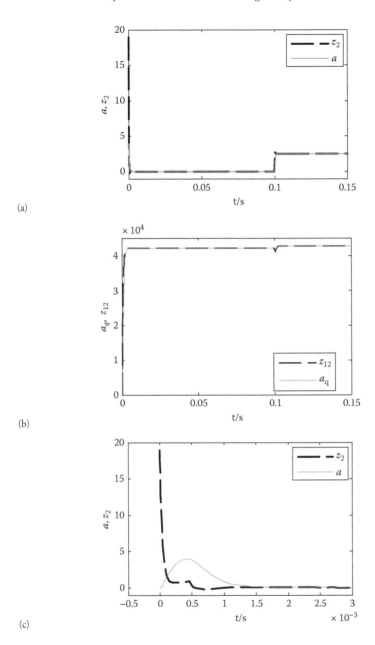

(a)

(b)

(c)

Figure 13.6 The actual and estimate value of disturbances under current ESOC method and: (a) a, z_2 of speed loop, (b) a_q, z_{12} of q-axis current loop, (c) the local curves of (a) in time domain [0,0.003] s, (d) the local curves of (b) in time domain [0,0.008] s, (e) the local curves of (a) in time domain [0.099,0.103] s, and (f) the local curves of (b) in time domain [0.099,0.106] s. *(Continued)*

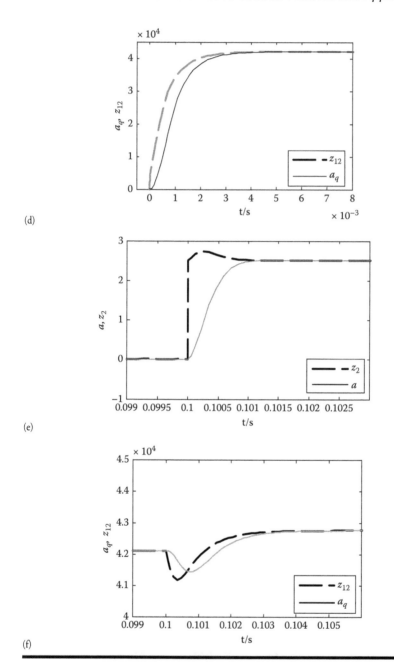

(d)

(e)

(f)

Figure 13.6 *(Continued)* **The actual and estimate value of disturbances under current ESOC method and: (a) a, z_2 of speed loop, (b) a_q, z_{12} of q-axis current loop, (c) the local curves of (a) in time domain [0,0.003] s, (d) the local curves of (b) in time domain [0,0.008] s, (e) the local curves of (a) in time domain [0.099,0.103] s, and (f) the local curves of (b) in time domain [0.099,0.106] s.**

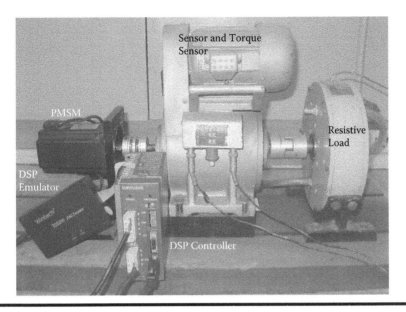

Figure 13.7   The experimental test setup.

Table 13.1   The performance comparisons.

Speed (rpm)	Algorithm	Overshoot (%)	Settling time (ms)	Antidisturbance comparison Decrease (rpm)	Increase (rpm)
	Speed ESOC	15.05	65	17	9.5
500	Current ESOC	10.85	63	12.75	7
	Speed ESOC	7.55	128	20	9.7
1000	Current ESOC	5.275	135	13.25	5.5
	Speed ESOC	5	195	22.5	11
1500	Current ESOC	3.5	208	14	5.5

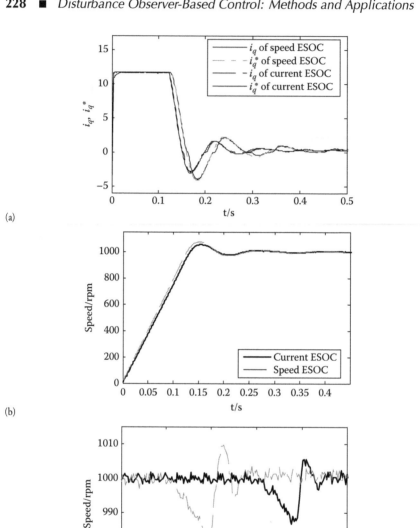

(a)

(b)

(c)

Figure 13.8 **Response curves of system when the reference speed is 1000 rpm.**
(a) i_q, i_q^*. **(b) Speed in the absence of load. (c) Speed in the presence of load.**

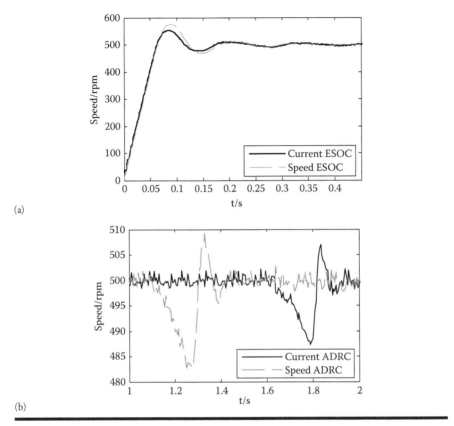

(a)

(b)

Figure 13.9   Speed response curves when the reference speed is 500 rpm. (a) In the absence of load. (b) In the presence of load.

double-pole of the current and speed loops are takes as $k_q = 25$, $p_q = 870$, and $k = 1.2$, $p = 330$, respectively.

The responses of PMSM system when the reference speed is 1000 rpm under the speed ESOC and current ESOC schemes are shown in Figure 13.8. It can be seen that for the current ESOC control scheme, it takes a shorter settling time for the q-axis current i_q to track its reference value i_q^*, for the speed output ω to reach the reference speed ω^*. In the presence of load disturbances, the maximum fluctuations of the system under the speed ESOC and current ESOC schemes are 20 rpm and 13.25 rpm, respectively. In addition, the speed response comparisons when the reference speed are 500 rpm and 1500 rpm are also given in Figures 13.9 and 13.10. A detail quantitative comparisons of performance can be found in Table 13.1.

The simulation and experiment results show that the proposed method can reduce the overshoot and improve the disturbance rejection property for PMSM system.

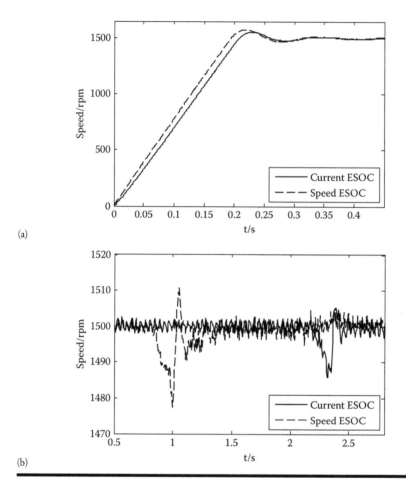

(a)

(b)

Figure 13.10 Speed response curves when the reference speed is 1500 rpm. (a) In the absence of load. (b) In the presence of load.

13.5 Summary

The speed-regulation problem for permanent magnet synchronous motor system has been addressed in this chapter. To improve the disturbance rejection ability, a control scheme which employs disturbance rejection control laws for not only speed loop but also the q-axis current loop, has been proposed. Considering the dynamics of q-axis current, the coupling between rotor speed and d-axis current as well as the back electromotive force have been regarded as lumped disturbances for the q-axis current loop. The lumped disturbances have been estimated by using an extended state observer. Thus, a composite control law consisting of proportional feedback and feedforward compensation based on disturbance estimation has been developed to control the q-axis current. Simulation and experiment comparisons have been presented to verify the effectiveness of the proposed method.

APPLICATION TO FLIGHT CONTROL SYSTEMS

VI

Chapter 14

Disturbance Rejection for Small-Scale Helicopters

14.1 Introduction

Small-scale helicopters are very attractive for a wide range of civilian and military applications due to their unique features. However, autonomous flight for small helicopters is quite challenging because they are naturally unstable, have strong nonlinearities and couplings, and are very susceptible to wind and small structural variations. Many control techniques have been applied to address the autonomous flight of helicopters including feedback linearization-based control [187], multivariable adaptive control [188], neural network adaptive control [189], state-dependent Riccati equation (SDRE) control [190], etc.

Recently, model-predictive control (MPC) has been recognized as a promising method in the unmanned aerial vehicle (UAV) community. The essential procedure in the implementation of MPC algorithms is to solve the formulated optimization problem (OP). For nonlinear system, MPC technique generally requires solving an optimization problem numerically at every sampling instant, which poses obstacles on the real-time implementation due to the heavy computational burden. Although the development of the avionics and microprocessor technology makes the online optimization possible, the implementation of computationally demanding nonlinear MPC on small UAVs is very challenging. The associated low bandwidth and computational delay make it very difficult to meet the control requirement for systems with fast dynamics such as helicopters. Only a few applications on helicopter flight control have been reported in [191, 192], where the authors adopt a high-level MPC to solve the tracking problem and rely on a local linear feedback controller to compensate the high-level MPC. Moreover, the formulated nonlinear optimization problem has to be solved by a secondary flight computer. The extra payload and power consumption are unsuitable for a small-scale helicopter.

To avoid online optimization, this chapter introduces an explicit nonlinear MPC (ENMPC) for trajectory tracking of autonomous helicopters. By approximating the tracking error and control efforts in the receding horizon using their Taylor expansion to a specified order, an analytic solution to nonlinear MPC can be found and consequently the closed-form controller can be formulated without online optimization [6]. The benefits of using this MPC algorithm are not only the elimination of the online optimization and the associated resource, but also a higher control bandwidth, which is very important for helicopters in aggressive flight scenarios.

Apart from the control method, there are practical issues in controlling autonomous helicopters from an engineering point of view. It is known that the control performance of MPC, or other model-based control technologies, heavily relies on the quality of the model. However, the model of high accuracy for a helicopter is difficult to obtain due to the complicated aerodynamic nature of the rotor system. On the other hand, due to the light-weighted structure, small-scale helicopters are more likely to be affected by wind gusts and other disturbances than their full size counterpart, and the physical parameters such as mass and inertia of moments can be easily altered by changing the payload and even its location. All these factors compromise the actual performance of the controller designed based on the nominal model.

To enhance the performance of ENMPC in a complex operation environment, a DO has been applied to estimate unknown disturbances in the control process [2, 40]. As the estimation of disturbances is provided, the control system can explicitly take them into account and compensate them. The advantage of the DOBC is that it preserves the tracking and other properties of the original baseline control while being able to compensate disturbances rather than resorting to a different control strategy.

To design a DO-augmented ENMPC for trajectory tracking of autonomous helicopters, two problems need to be addressed, namely, designing the nonlinear disturbance observer to estimate the disturbances acting on the helicopter, integrating the disturbance information into ENMPC to compensate their influences. To this end, another contribution of this chapter lies in the synthesis of the ENMPC and DO by exploiting the helicopter model structure. The disturbances are assumed to exist in certain channels of the helicopter where the coupling terms can also be lumped into disturbance terms. In this way an ENMPC is derived under the assumption that all the disturbances are measurable and then these disturbances are replaced by their estimation provided by the proposed disturbance observers. On the other hand, the lumped disturbance terms simplify the model structure, allowing the derivation of ENMPC for helicopters. The composite control framework provides a promising solution to autonomous helicopter trajectory tracking in the presence of uncertainties and disturbances. The performance of the proposed control system is tested through

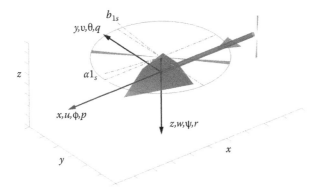

Figure 14.1 **Helicopter frame.**

simulations and verified in our indoor flight testbed. The contents of this chapter mainly refer to [193].

14.2 Helicopter Modeling

A helicopter is a highly nonlinear system with multiple-inputs–multiple-outputs (MIMO) and complex internal couplings. The complete model taking into account the flexibility of the rotors and fuselage usually results in a model of high degrees-of-freedom. The complexity of such a model would make the following system identification much more difficult. A practical way to deal with this issue is to capture the primary dynamics by a simplified model and treat the other trivial factors that affect dynamics as uncertainties or disturbances. The general dynamics of a small-scale helicopter can be captured by a six-degrees-of-freedom rigid-body model augmented with a simplified rotor dynamic model [194, 195], as shown in Figure 14.1. In this way, the kinematics of the helicopter, i.e., the position and the orientation represented by Z-Y-X Euler angles, can be expressed as:

$$[\dot{x} \ \dot{y} \ \dot{z}]^T = \mathbf{R}_b^i(\phi, \theta, \psi)[u \ v \ w]^T \tag{14.1}$$

$$\begin{bmatrix} \dot{\phi} \\ \dot{\theta} \\ \dot{\psi} \end{bmatrix} = \begin{bmatrix} 1 & \sin\phi \tan\theta & \cos\phi \tan\phi \\ 0 & \cos\phi & -\sin\phi \\ 0 & \sin\phi \sec\theta & \cos\phi \sec\theta \end{bmatrix} \begin{bmatrix} p \\ q \\ r \end{bmatrix} \tag{14.2}$$

where (x, y, z) describe the helicopter inertial position, (u, v, w) are velocities along three body axes, (p, q, r) are angular rates and (ϕ, θ, ψ) are attitude angles, and $\mathbf{R}_b^i$ is the transformation matrix from body to inertial coordinates given in (14.3) with short notation c for cosine and s for sine.

$$\mathbf{R}_b^i(\phi, \theta, \psi) = \begin{bmatrix} c\theta c\psi & s\phi s\theta c\psi - c\phi s\psi & c\phi s\theta c\psi + s\phi s\psi \\ c\theta s\psi & s\phi s\theta s\psi + c\phi c\psi & c\phi s\theta s\psi - s\phi c\psi \\ -s\theta & s\phi c\theta & c\phi c\theta \end{bmatrix} \quad (14.3)$$

The model of translational dynamics of helicopters used in this chapter is modified by keeping the thrust of main rotor as a dominating force and considering other force contributions as disturbances, such that

$$\dot{u} = vr - wq - g\sin\theta + d_x$$

$$\dot{v} = wp - ur + g\cos\theta\sin\phi + d_y \quad (14.4)$$

$$\dot{w} = uq - vp + g\cos\theta\cos\phi + T + d_z$$

where, T is the normalized main rotor thrust controlled by collective pitch δ_{col}, as $T = g + Z_w w + Z_{col}\delta_{col}$, and (d_x, d_y, d_z) are normalized force disturbances that include external wind gusts, internal couplings, and unmodeled dynamics. These force disturbances directly affect the translational dynamics and result in tracking error. As force disturbances are not in the channels of control inputs, they are called mismatched disturbances. This modification, on one hand, increases the valid range of the model compared to simplified helicopter models for control design that neglect all other forces other than the main thrust [187, 196, 197]. On the other hand, it reduces the workload of deriving the ENMPC for helicopters as different forces are lumped into one term .

In rotation dynamics, the torques are generated by the tilting of main rotor such that

$$\dot{p} = -qr(I_{yy} - I_{zz})/I_{xx} + L_a a + L_b b$$

$$\dot{q} = -pr(I_{zz} - I_{xx})/I_{yy} + M_a a + M_b b \quad (14.5)$$

$$\dot{r} = -pq(I_{xx} - I_{yy})/I_{zz} + N_r r + N_{col}\delta_{col} + N_{ped}\delta_{ped}$$

where a and b are flapping angles to depict the tilt of the main rotor along the longitudinal and lateral axis, respectively; the other parameters in the model are the stability and control derivatives, whose values for the helicopter used in this study are obtained by system identification. The flapping angles a and b of the main rotor are originally controlled by lateral and longitudinal cyclic δ_{lat} and δ_{lon}.

Their relationship can be approximated by steady state dynamics of the main rotor [190]:

$$a = -\tau q + A_{lat}\delta_{lat} + A_{lon}\delta_{lon}$$
$$b = -\tau p + B_{lat}\delta_{lat} + B_{lon}\delta_{lon}$$

(14.6)

Apart from force disturbances, small-scale helicopters also subject to structural uncertainties and are vulnerable to physical alterations like payload change. These factors are commonly ignored in the control design, as they can be compensated by setting control trims in the implementation. To save the trim tuning process in the real life operation, we consider trims errors in the control channel as disturbances. Thereby, combining (14.5) and (14.6) yields

$$\dot{p} = -L_{pq} + L_{lat}(\delta_{lat} + d_{lat}) + L_{lon}(\delta_{lon} + d_{lon})$$
$$\dot{q} = -M_{pq} + M_{lat}(\delta_{lat} + d_{lat}) + M_{lon}(\delta_{lon} + d_{lat})$$
$$\dot{r} = -N_{pr} + N_r r + N_{col}\delta_{col} + N_{ped}(\delta_{ped} + d_{ped})$$

(14.7)

where

$$L_{pq} = qr(I_{yy} - I_{zz})/I_{xx} + \tau(L_a q + L_b p),$$
$$M_{pq} = pr(I_{zz} - I_{xx})/I_{yy} + \tau(M_a q + M_b p),$$
$$N_{pq} = pq(I_{xx} - I_{yy})/I_{zz},$$

(14.8)

$$L_{lat} = L_a A_{lat} + L_b B_{lat}, \qquad M_{lat} = M_a A_{lat} + M_b B_{lat},$$
$$L_{lon} = L_a A_{lon} + L_b B_{lon}, \qquad M_{lon} = M_a A_{lon} + M_b B_{lon},$$

and d_{lat}, d_{lon}, and d_{ped} account for different trim errors. In addition, since they are combined into the angular dynamics and affect the angular rate directly, they can be considered as torque disturbances.

The modified helicopter model by combining (14.1)–(14.7) can be expressed by a general affine form:

$$\dot{x} = f(x) + g_1(x)u + g_2(x)d$$
$$y = h(x)$$

(14.9)

where $x = [\, x\ y\ z\ u\ v\ w\ p\ q\ r\ \phi\ \theta\ \psi\,]^T$ is the helicopter state, y is the output of the helicopter, and $d = [\, d_x\ d_y\ d_z\ d_{lat}\ d_{lon}\ d_{ped}\,]^T$ is the lumped disturbance acting on the helicopter. In the trajectory tracking control of an autonomous helicopter, the interested outputs are the position and heading angle. Thus, $y = [\, x\ y\ z\ \psi\,]^T$.

14.3 Explicit Nonlinear MPC With Disturbances

Trajectory tracking is the basic function required when an autonomous helicopter performs a task. To this end, we need to design a controller such that the output $\mathbf{y}(t)$ of the helicopter (14.9) tracks the prescribed reference $\mathbf{w}(t)$. In the MPC strategy, tracking control can be achieved by minimising a receding horizon performance index

$$J = \frac{1}{2} \int_0^T (\hat{\mathbf{y}}(t+\tau) - \mathbf{w}(t+\tau))^T Q(\hat{\mathbf{y}}(t+\tau) - \mathbf{w}(t+\tau)) d\tau \qquad (14.10)$$

where weighting matrix $Q = \mathrm{diag}\{q_1, q_2, q_3, q_4\}$, $q_i > 0$, $i = 1, 2, 3, 4$. Note that the hatted variables belong to the prediction time frame.

Conventional MPC algorithm requires solving of an optimisation problem at every sampling instant to obtain the control signals. To avoid the computationally intensive online optimisation, we adopt an explicit solution for the nonlinear MPC problem based on the approximation of the tracking error in the receding prediction horizon [6].

14.3.1 Output Approximation

For a nonlinear MIMO system like the helicopter, it is well known that after differentiating the outputs for a specific number of times, the control inputs appear in the expressions. The number of times of differentiation is defined as *relative degree*. For the helicopter with output $\mathbf{y} = [\,x\ y\ z\ \psi\,]'$ and the corresponding input $\mathbf{u} = [\,\delta_{lon}\ \delta_{lat}\ \delta_{col}\ \delta_{ped}\,]$, the relative degree is a vector, $\rho = [\,\rho_1\ \rho_2\ \rho_3\ \rho_4\,]$. If continuously differentiating the output after the control input appears, the derivatives of control input appear, where the number of the input derivatives r is defined as the *control order*.

Since the helicopter model has different relative degrees, the control order r is first specified in the controller design. The ith output of the helicopter in the receding horizon can be approximated by its Taylor series expansion up to order $\rho_i + r$:

$$\hat{y}_i(t+\tau) \approx y_i(t) + \tau \dot{y}_i(t) + \ldots + \frac{\tau^{r+\rho_i}}{(r+\rho_i)!} y_i^{[r+\rho_i]}(t)$$

$$= \begin{bmatrix} 1 & \tau & \ldots & \frac{\tau^{r+\rho_i}}{(r+\rho_i)!} \end{bmatrix} \begin{bmatrix} y_i(t) \\ \dot{y}_i(t) \\ \ldots \\ y_i^{[r+\rho_i]}(t) \end{bmatrix}, 0 \leq \tau \leq T \qquad (14.11)$$

where $i = 1, 2, 3, 4$. In this way, the approximation of the overall output of the helicopter can be cast in a matrix form:

$$\hat{\mathbf{y}}(t + \tau) = \begin{bmatrix} \hat{x}(t + \tau) \\ \hat{y}(t + \tau) \\ \hat{z}(t + \tau) \\ \hat{\psi}(t + \tau) \end{bmatrix} = \begin{bmatrix} \hat{y}_1(t + \tau) \\ \hat{y}_2(t + \tau) \\ \hat{y}_3(t + \tau) \\ \hat{y}_4(t + \tau) \end{bmatrix}$$

$$= \begin{bmatrix} 1, \tau, \ldots, \frac{\tau^{r+\rho_1}}{(r+\rho_1)!} & \cdots & 0_{1\times(r+\rho_4+1)} \\ \cdots & \cdots & \cdots \\ 0_{1\times(r+\rho_1+1)} & \cdots & 1, \tau, \ldots, \frac{\tau^{r+\rho_4}}{(r+\rho_4)!} \end{bmatrix} \begin{bmatrix} y_1(t) \\ \dot{y}_1(t) \\ \vdots \\ y_1^{[r+\rho_1]}(t) \\ \vdots \\ y_4(t) \\ \dot{y}_4(t) \\ \vdots \\ y_4^{[r+\rho_4]}(t) \end{bmatrix} \qquad (14.12)$$

For each channel in the output matrix, the control orders r are the same and can be decided during the control design, whereas the relative degrees ρ_i are different but determined by the helicopter model structure. Manipulating the output matrix (14.12) gives the following partition:

$$\hat{\mathbf{y}}(t + \tau) = \begin{bmatrix} \bar{\tau}_1 & \cdots & 0_{1\times\rho_4} & | & & & \\ \cdots & \cdots & \cdots & | & \bar{\tau}_1 & \cdots & \bar{\tau}_{r+1} \\ 0_{1\times\rho_1} & \cdots & \bar{\tau}_4 & | & & & \end{bmatrix} \qquad (14.13)$$

$$\begin{bmatrix} [\bar{Y}_1(t)^T & \cdots & \bar{Y}_4(t)^T | \bar{Y}_1(t)^T & \cdots & \bar{Y}_r(t)^T] \end{bmatrix}^T$$

where

$$\bar{Y}_i = \begin{bmatrix} y_i(t) & \dot{y}_i(t) & \cdots & y_i^{[\rho_i-1]} \end{bmatrix}^T, i = 1, 2, 3, 4 \qquad (14.14)$$

$$\bar{Y}_i = \begin{bmatrix} y_1^{[\rho_1+i-1]} & y_2^{[\rho_2+i-1]} & \cdots & y_4^{[\rho_4+i-1]} \end{bmatrix}^T, i = 1, \ldots, r+1 \qquad (14.15)$$

$$\bar{\tau}_i = \begin{bmatrix} 1 & \tau & \cdots & \frac{\tau^{\rho_i-1}}{(\rho_i-1)!} \end{bmatrix}, i = 1, 2, 3, 4 \qquad (14.16)$$

and

$$\bar{\tau} = \text{diag} \left\{ \frac{\tau^{\rho_1+i-1}}{(\rho_1+i-1)!} \cdots \frac{\tau^{\rho_4+i-1}}{(\rho_4+i-1)!} \right\} \tag{14.17}$$

It can be observed from Equation (14.13) that the prediction of the helicopter output $\hat{\mathbf{y}}(t+\tau)$, $0 \leq \tau \leq T$, in the receding horizon needs the derivatives of each output of the helicopter up to $r + \rho_i$ order at time instant t. Except for the output $\mathbf{y}(t)$ itself that can be directly measured, the other derivatives have to be derived according to the helicopter model (14.9). During this process the control input will appear in the ρ_ith derivatives, where $i = 1, 2, 3, 4$.

The first derivatives can be obtained from the helicopter's kinematics model:

$$\begin{bmatrix} \dot{y}_1 \\ \dot{y}_2 \\ \dot{y}_3 \end{bmatrix} = \begin{bmatrix} \dot{x} \\ \dot{y} \\ \dot{z} \end{bmatrix} = \mathbf{R}_b^i \cdot \begin{bmatrix} u \\ v \\ w \end{bmatrix} \tag{14.18}$$

$$\dot{y}_4 = \dot{\psi} = q \sin \phi \sec \theta + r \cos \phi \sec \theta \tag{14.19}$$

Differentiating (14.18) and (14.19) with substitution of helicopter dynamics (14.1) yields the second derivatives:

$$\begin{bmatrix} \ddot{y}_1 \\ \ddot{y}_2 \\ \ddot{y}_3 \end{bmatrix} = \begin{bmatrix} \ddot{x} \\ \ddot{y} \\ \ddot{z} \end{bmatrix} = \mathbf{R}_b^i \begin{bmatrix} d_x \\ d_y \\ T + d_z \end{bmatrix} + \begin{bmatrix} 0 \\ 0 \\ g \end{bmatrix}, \tag{14.20}$$

where $T = Z_w w + Z_{col}\delta_{col} - g$ is the normalized main rotor thrust, and

$$\ddot{y}_4 = \ddot{\psi} = q \frac{\cos \phi}{\cos \theta} \dot{\phi} + q \frac{\sin \phi \sin \theta}{\cos^2 \theta} \dot{\theta} - r \frac{\sin \phi}{\cos \theta} \dot{\phi} + r \frac{\cos \phi \sin \theta}{\cos^2 \theta} \dot{\theta} - L_{pq} \frac{\sin \phi}{\cos \theta} + N_r \frac{\cos \phi}{\cos \theta} r$$

$$+ L_{lat} \frac{\sin \phi}{\cos \theta} (\delta_{lat} + d_{lat}) + L_{lon} \frac{\sin \phi}{\cos \theta} (\delta_{lon} + d_{lon}) + N_{col} \frac{\cos \phi}{\cos \theta} \delta_{col}$$

$$+ N_{ped} \frac{\cos \phi}{\cos \theta} (\delta_{ped} + d_{ped}) \tag{14.21}$$

Note that although control input δ_{col} appears in (14.20), the other control inputs do not, so we have to continue differentiating the first three outputs. To facilitate the derivation, we adopt the relationship $\dot{\mathbf{R}}_b^i = \mathbf{R}_b^i \hat{\omega}$ by using skew-symmetric matrix $\hat{\omega} \in \mathbb{R}^{3 \times 3}$:

$$\hat{\omega} = \begin{bmatrix} 0 & -r & q \\ r & 0 & -p \\ -q & p & 0 \end{bmatrix}. \tag{14.22}$$

Thus, the third and fourth derivatives of the position output can be written as:

$$
\begin{bmatrix} y_1^{[3]} \\ y_2^{[3]} \\ y_3^{[3]} \end{bmatrix} = \begin{bmatrix} x^{[3]} \\ y^{[3]} \\ z^{[3]} \end{bmatrix} = \mathbf{R}_b^i \hat{\omega} \begin{bmatrix} d_x \\ d_y \\ T + d_z \end{bmatrix} + \mathbf{R}_b^i \begin{bmatrix} 0 \\ 0 \\ Z_w \dot{w} + Z_{col} \dot{\delta}_{col} \end{bmatrix}, \qquad (14.23)
$$

and

$$
\begin{bmatrix} y_1^{[4]} \\ y_2^{[4]} \\ y_3^{[4]} \end{bmatrix} = \begin{bmatrix} x^{[4]} \\ y^{[4]} \\ z^{[4]} \end{bmatrix} = \mathbf{R}_b^i \hat{\omega}\hat{\omega} \begin{bmatrix} d_x \\ d_y \\ T + d_z \end{bmatrix} + 2\mathbf{R}_b^i \hat{\omega} \begin{bmatrix} 0 \\ 0 \\ Z_w \dot{w} + Z_{col} \dot{\delta}_{col} \end{bmatrix} +
$$

$$
\mathbf{R}_b^i \begin{bmatrix} -N_r r d_y - M_{pq}(T + d_z) \\ N_r r d_x + L_{pq}(T + d_z) \\ M_{pq} d_x - L_{pq} d_y + Z_w \dot{w} \end{bmatrix} +
$$

$$
\overline{A}(\mathbf{x}, \mathbf{d}) \begin{bmatrix} \delta_{lat} + d_{lat} & \delta_{lon} + d_{lon} & \ddot{\delta}_{col} & \delta_{ped} + d_{ped} \end{bmatrix}^T \qquad (14.24)
$$

where

$$
\overline{A}(\mathbf{x}, \mathbf{d}) = \mathbf{R}_b^i \begin{bmatrix} M_{lat}(T + d_z) & M_{lon}(T + d_z) & 0 & -N_{ped} d_y \\ -L_{lat}(T + d_z) & -L_{lon}(T + d_z) & 0 & N_{ped} d_x \\ -M_{lat} d_x + L_{lat} d_y & -M_{lon} d_x + L_{lon} d_y & Z_{col} & 0 \end{bmatrix}
$$

$$(14.25)$$

At this stage, the control inputs explicitly appear in (14.24). Therefore, the vector relative degree for the helicopter is $\rho = [\, 4 \ 4 \ 4 \ 2 \,]$. Note that in the formulation of (14.24) $\ddot{\delta}_{col}$ is the new control input, whereas δ_{col} and $\dot{\delta}_{col}$ are treated as the states which can be obtained by adding integrators. This procedure is known as achieving relative degree through dynamics extension [106].

By invoking (14.18)–(14.23), we now can construct matrix $\bar{Y}_i$, $i = 1, 2, 3, 4$. However, in order to find the elements in $\bar{Y}_i$, $i = 1, 2, \ldots, r + 1$, further manipulation is required. By combining (14.21) and (14.24) and utilizing the Lie notation [106], we have:

$$
\bar{Y}_1 = \begin{bmatrix} y_1^{[\rho_1]} \\ y_2^{[\rho_2]} \\ y_3^{[\rho_3]} \\ y_4^{[\rho_4]} \end{bmatrix} = \begin{bmatrix} x^{[4]} \\ y^{[4]} \\ z^{[4]} \\ \psi^{[2]} \end{bmatrix} = \begin{bmatrix} L_f^{\rho_1} h_1(\mathbf{x}, \mathbf{d}) \\ L_f^{\rho_2} h_2(\mathbf{x}, \mathbf{d}) \\ L_f^{\rho_3} h_3(\mathbf{x}, \mathbf{d}) \\ L_f^{\rho_4} h_4(\mathbf{x}, \mathbf{d}) \end{bmatrix} + A(\mathbf{x}, \mathbf{d}) \bar{u} \qquad (14.26)
$$

where $\bar{\mathbf{u}} = [\,\delta_{lat} + d_{lat}\ \ \delta_{lon} + d_{lon}\ \ \ddot{\delta}_{col}\ \ \delta_{ped} + d_{ped}\,]$; nonlinear terms $L_f^{\rho_i} h_i(\mathbf{x}, \mathbf{d})$, $i = 1, 2, 3, 4$, can be found in the previous derivation, and

$$A(\mathbf{x}, \mathbf{d}) = \begin{bmatrix} \overline{A}(\mathbf{x}, \mathbf{d}) \\ \underline{A}(\mathbf{x}, \mathbf{d}) \end{bmatrix}. \tag{14.27}$$

where $\overline{A}(\mathbf{x}, \mathbf{d})$ is given in Equation (14.25) and

$$\underline{A}(\mathbf{x}, \mathbf{d}) = \begin{bmatrix} L_{lat} \dfrac{\sin\phi}{\cos\theta} & L_{lon} \dfrac{\sin\phi}{\cos\theta} & 0 & N_{ped} \dfrac{\cos\phi}{\cos\theta} \end{bmatrix}. \tag{14.28}$$

Differentiating (14.26) with respect to time together with substitution of the system's dynamics gives

$$\bar{Y}_2 = \begin{bmatrix} y_1^{[\rho_1+1]} \\ y_2^{[\rho_2+1]} \\ y_3^{[\rho_3+1]} \\ y_4^{[\rho_4+1]} \end{bmatrix} = \begin{bmatrix} L_f^{\rho_1+1} h_1(\mathbf{x}) \\ L_f^{\rho_2+1} h_2(\mathbf{x}) \\ L_f^{\rho_3+1} h_3(\mathbf{x}) \\ L_f^{\rho_4+1} h_4(\mathbf{x}) \end{bmatrix} + A(\mathbf{x}, \mathbf{d})\bar{\mathbf{u}}^{[1]} + p_1(\mathbf{x}, \bar{\mathbf{u}}) \tag{14.29}$$

where $p_1(\mathbf{x}, \bar{\mathbf{u}})$ is a nonlinear vector function of $\mathbf{x}$ and $\bar{\mathbf{u}}$. By repeating this procedure, the higher derivatives of the output and $\bar{Y}_i$, $i = 1, 2, \ldots, r$, can be calculated and finally we have

$$\bar{Y}_{r+1} = \begin{bmatrix} y_1^{[\rho_1+r]} \\ y_2^{[\rho_2+r]} \\ y_3^{[\rho_3+r]} \\ y_4^{[\rho_4+r]} \end{bmatrix} = \begin{bmatrix} L_f^{\rho_1+r} h_1(\mathbf{x}) \\ L_f^{\rho_2+r} h_2(\mathbf{x}) \\ L_f^{\rho_3+r} h_3(\mathbf{x}) \\ L_f^{\rho_4+r} h_4(\mathbf{x}) \end{bmatrix} + A(\mathbf{x}, \mathbf{d})\bar{\mathbf{u}}^{[r]} + p_r(\mathbf{x}, \bar{\mathbf{u}}, \bar{\mathbf{u}}^{[1]}, \ldots, \bar{\mathbf{u}}^{[r]})$$

$$\tag{14.30}$$

So far by exploiting the helicopter model, the elements to construct Y and $\bar{Y}$ in Equation (14.13) are available. Therefore, the output of the helicopter in the future horizon $\mathbf{y}(t+\tau)$ can be expressed by its Taylor expansion in a generalized linear form with respect to the prediction time τ and current states as shown in Equation (14.13).

In the same fashion as in Equation (14.13), the reference in the receding horizon $w(t + \tau)$, $0 \le \tau \le T$ can also be approximated by:

$$\mathbf{w}(t + \tau) = \begin{bmatrix} w_1(t + \tau) \\ w_2(t + \tau) \\ w_3(t + \tau) \\ w_4(t + \tau) \end{bmatrix} = \begin{bmatrix} T_f & T_s \end{bmatrix} \begin{bmatrix} \bar{W}_1(t)^T \ldots \bar{W}_4(t)^T | \bar{W}_1(t)^T \ldots \bar{W}_{r+1}(t)^T \end{bmatrix}^T$$

$$\tag{14.31}$$

where

$$
T_f = \begin{bmatrix} \bar{\tau}_1 & \cdots & 0_{1 \times \rho_4} \\ \vdots & \ddots & \vdots \\ 0_{1 \times \rho_1} & \cdots & \bar{\tau}_4 \end{bmatrix} \tag{14.32}
$$

and

$$
T_s = \begin{bmatrix} \bar{\tau}_1 & \cdots & \bar{\tau}_{r+1} \end{bmatrix} \tag{14.33}
$$

and the construction of $\bar{W}_i(t)$, $i = 1, 2, 3, 4$, and $\tilde{W}_i$, $i = 1, \ldots, r + 1$, can refer to the structure of $\bar{Y}_i(t)$ and $\tilde{Y}_i$, respectively.

14.3.2 Explicit Nonlinear MPC Solution

The conventional MPC needs to solve a formulated optimisation problem to generate the control signal, where the control performance index is minimized with respect to the future control input over the prediction horizon. After the output is approximated by its Taylor expansion, the control profile can be defined as

$$
\bar{\mathbf{u}}(t + \tau) = \bar{\mathbf{u}}(t) + \tau \bar{\mathbf{u}}^{[1]}(t) + \ldots + \frac{\tau^r}{r!} \bar{\mathbf{u}}^{[r]}(t), \ 0 \le \tau \le T \tag{14.34}
$$

Thereby, the helicopter outputs depend on the control variables $\bar{\mathbf{u}} = \{\bar{\mathbf{u}}, \bar{\mathbf{u}}^{[1]}, \ldots, \bar{\mathbf{u}}^{[r]}\}$.

Recalling the performance index (14.10) and the output reference approximation (14.13) and (14.31), we have:

$$
J = \frac{1}{2} (\bar{Y}(t) - \bar{W}(t))^T \begin{bmatrix} T_1 & T_2 \\ T_2^T & T_3 \end{bmatrix} (\bar{Y}(t) - \bar{W}(t)) \tag{14.35}
$$

where

$$
\bar{Y}(t) = \begin{bmatrix} \bar{Y}_1(t)^T & \cdots & \bar{Y}_4(t)^T \mid \tilde{Y}_1(t)^T & \cdots & \tilde{Y}_r(t)^T \end{bmatrix}^T, \tag{14.36}
$$

$$
\bar{W}(t) = \begin{bmatrix} \bar{W}_1(t)^T & \cdots & \bar{W}_4(t)^T \mid \tilde{W}_1(t)^T & \cdots & \tilde{W}_{r+1}(t)^T \end{bmatrix}^T, \tag{14.37}
$$

$$
T_1 = \int_0^T T_f^T Q T_f d\tau, \tag{14.38}
$$

$$
T_2 = \int_0^T T_f^T Q T_s d\tau, \tag{14.39}
$$

and

$$T_3 = \int_0^T T_s^T Q T_s \, d\tau. \tag{14.40}$$

Therefore, instead of minimising the performance index (14.10) with respect to control profile $\mathbf{u}(t + \tau)$, $0 < \tau < T$ directly, we can minimize the approximated index (14.35) with respect to $\bar{\mathbf{u}}$, where the necessary condition for the optimality is given by

$$\frac{\partial J}{\partial \bar{\mathbf{u}}} = 0 \tag{14.41}$$

After solving the nonlinear Equation (14.41), we can obtain the optimal control variables $\bar{\mathbf{u}}^*$ to construct the optimal control profile defined by Equation (14.34). As in MPC only the current control in the control profile is implemented, the explicit solution is $\bar{\mathbf{u}}^* = \bar{\mathbf{u}}(t + \tau)$, for $\tau = 0$. The resulting controller is given by

$$\bar{\mathbf{u}}^* = -A(\mathbf{x}, \mathbf{d})^{-1}(K M_\rho + M_1) \tag{14.42}$$

where $K \in \mathbb{R}^{4 \times (\rho_1 + \dots + \rho_4)}$ is the first four rows of the matrix $T_3^{-1} T_2^T \in \mathbb{R}^{4(r+1) \times (\rho_1 + \dots + \rho_4)}$ where the ijth block of T_2 is of $\rho_i \times 4$ matrix, and all its elements are zeros except the ith column is given by

$$\left[q_i \frac{T^{\rho_i+j}}{(\rho_i+j-1)!(\rho_i+j)} \quad \cdots \quad q_i \frac{T^{2\rho_i+j-1}}{(\rho_i+j-1)!(\rho_i-1)!(2\rho_i+j-1)} \right]^T \tag{14.43}$$

for $i = 1, 2, 3, 4$ and $j = 1, 2, \dots, r+1$, and ijth block of T_3 is given by

$$\mathrm{diag}\left\{ q_1 \frac{T^{2\rho_1+i+j-1}}{(\rho_1+i-1)!(\rho_1+j-1)!(2\rho_1+i+j-1)}, \cdots, q_4 \frac{T^{2\rho_4+i+j-1}}{(\rho_4+i-1)!(\rho_4+j-1)!(2\rho_4+i+j-1)} \right\} \tag{14.44}$$

for $i, j = 1, 2, \dots, r+1$; the matrix $M_\rho \in \mathbb{R}^{\rho_1 + \dots + \rho_4}$ and matrix $M_i \in \mathbb{R}^4$ are defined as:

$$M_\rho = \begin{bmatrix} \bar{Y}_1(t)^T \\ \vdots \\ \bar{Y}_4(t)^T \end{bmatrix} - \begin{bmatrix} \bar{W}_1(t)^T \\ \vdots \\ \bar{W}_4(t)^T \end{bmatrix} \tag{14.45}$$

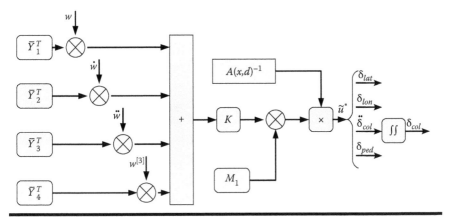

Figure 14.2 ENMPC structure.

and

$$
M_i = \begin{bmatrix} L_f^{\rho_1+i-1} h_1(t) \\ L_f^{\rho_2+i-1} h_2(t) \\ \vdots \\ L_f^{\rho_4+i-1} h_4(t) \end{bmatrix} - \bar{W}_i(t)^T, \; i = 1, 2, \ldots, r+1. \tag{14.46}
$$

The detailed derivation and closed-loop stability can refer to [6]. The overall controller structure is shown in Figure 14.2.

The system has trivial zero dynamics as $\rho_1 + \rho_2 + \rho_3 + \rho_4 = 14$, which is the order of the helicopter dynamics plus the dynamic extension. If the disturbance terms are set to zero, the controller is equivalent to that designed using the nominal model. The information of disturbances are hold in the controller to eliminate their influences.

In order to implement the above control strategy, the disturbances must be available, which is unrealistic for helicopter flight. The next section will introduce a nonlinear disturbance observer to estimate these unavailable disturbances.

14.4 Disturbance Observer-Based Control

14.4.1 Disturbance Observer

For a system like a small-scale helicopter, precisely modelling its dynamics or directly measuring the disturbances acting on it is very difficult. However, the disturbance observer technique provides an alternative way to estimate them. In this section, we introduce a nonlinear disturbance observer to estimate the lumped unknown disturbances d in the general form of helicopter model (14.9). The disturbance observer is given by [5],

$$\hat{d} = z + p(x)$$

$$\dot{z} = -l(x)g_2(x)z - l(x)[g_2(x)p(x) + f(x) + g_1(x)u]$$

(14.47)

where $\hat{d} = \begin{bmatrix} \hat{d}_x & \hat{d}_y & \hat{d}_z & \hat{d}_{lat} & \hat{d}_{lon} & \hat{d}_{ped} \end{bmatrix}$ is the estimation of disturbances; z is the internal state of the nonlinear observer, $p(x)$ is a nonlinear function to be designed, and $l(x)$ is the nonlinear observer gain given by

$$l(x) = \frac{\partial p(x)}{\partial x}$$

(14.48)

In this observer, the estimation error is defined as $e_d = d - \hat{d}$. Under the assumption that the disturbance is slowly varying compared with the observer dynamics and by combining Equation (14.47), Equation (14.48) and Equation (14.9), it can be shown that the estimation error has the following property:

$$\dot{e}_d = \dot{d} - \dot{\hat{d}} = -\dot{z} - \frac{\partial p(x)}{\partial x}\dot{x} = -l(x)g_2(x)e_d$$

(14.49)

Therefore, $\hat{d}(t)$ approaches $d(t)$ exponentially if $p(x)$ is chosen such that Equation (14.49) is globally exponentially stable for all $x \in R^n$.

The design of a disturbance observer essentially is to choose an appropriate gain $l(x)$ and associated $p(x)$ such that the convergence of estimation error is guaranteed. Thereby, there exists a considerable degree of freedom. Since the disturbance input matrix $g_2(x)$ for the helicopter model is a constant matrix as:

$$g_2(x) = \begin{bmatrix} 0_{3\times3} & & & 0_{3\times3} \\ 1 & 0 & 0 & \\ 0 & 1 & 0 & 0_{3\times3} \\ 0 & 0 & 1 & \\ & & & L_{lat} & L_{lon} & 0 \\ & 0_{3\times3} & M_{lat} & M_{lon} & 0 \\ & & & 0 & 0 & N_{ped} \\ 0_{3\times3} & & & 0_{3\times3} \end{bmatrix},$$

(14.50)

we can choose $l(x)$ as a constant matrix such that all the eigenvalues of matrix $-l(x)g_2(x)$ have negative real part. Next, integrating $l(x)$ with respect to the helicopter state x yields $p(x) = l(x)x$. The observer gain matrix $l(x)$ corresponding to

g_2 is designed in the form:

$$l(x) = \begin{bmatrix} 0_{3\times3} & L_1 & 0_{3\times3} & 0_{3\times3} \\ 0_{3\times3} & 0_{3\times3} & L_2 & 0_{3\times3} \end{bmatrix} \tag{14.51}$$

where matrix $L_1 = \text{diag}\{l_1, l_2, l_3\}$ and

$$L_2 = \text{diag}\{l_4, l_5, l_6\} \begin{bmatrix} L_{lat} & L_{lon} & 0 \\ M_{lat} & M_{lon} & 0 \\ 0 & 0 & N_{ped} \end{bmatrix}^{-1} \tag{14.52}$$

for $l_i > 0$, $i = 1, \ldots 6$. Thereby, $-l(x)g(x) = -\text{diag}\{l_1, \ldots l_6\}$. From the above analysis, it can be seen that the convergence of the disturbance observer is guaranteed, regardless of the helicopter state $\mathbf{x}$.

14.4.2 Composite Controller

External force and torque disturbances generated by wind, turbulence, and other factors coupled with modelling errors and uncertainties may significantly degrade the helicopter tracking performance. It may even cause instability unless their influence has been properly taken into account in the system design. It can be noted that in the previous derivation of ENMPC, the lumped disturbances appear in the control law. Therefore, once the disturbance observer provides the estimation of disturbances, ENMPC controller takes into account the disturbances by replacing the disturbance by their estimation and achieves desired tracking performance. Let $\mathbf{d} = [\mathbf{d_f}\ \mathbf{d_e}]$, $\mathbf{d_f} = [d_x\ d_y\ d_z]$, and $\mathbf{d_e} = [d_{lat}\ d_{lon}\ d_{ped}]$. The composite controller law using the estimated disturbances is given in

$$\bar{\mathbf{u}} = -A(\mathbf{x}, \hat{\mathbf{d}}_f)^{-1}(K\hat{M}_\rho + \hat{M}_1) \tag{14.53}$$

where, the hatted variables denote the estimated values. If we consider trim errors in the helicopter dynamics, the overall control is in

$$\mathbf{u} = \bar{\mathbf{u}} - \hat{\mathbf{u}}_0 \tag{14.54}$$

where $\hat{\mathbf{u}}_0 = [\hat{d}_{lat}\ \hat{d}_{lon}\ 0\ \hat{d}_{ped}]^T$ is the control trim error estimated by the disturbance observer. The composite controller structure is illustrated in Figure 14.3.

14.5 Stability Analysis

The stabilities of the ENMPC and the disturbance observer are guaranteed in their design procedures outlines in Section 14.3 and 14.4, respectively. However, the

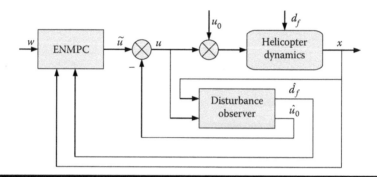

Figure 14.3 Composite controller structure.

stability of closed-loop system still needs to be examined, because the true disturbances are replaced by their estimation in composite controller (14.54), and there are interactions interactions among the ENMPC, the disturbance observer and the helicopter dynamics.

The closed-loop dynamics under the composite control law can be examined by applying Equation (14.54) into helicopter model (14.9). Since the resulting system is too complicated, we define a new coordinates to describe the closed-loop system. First, let position tracking error defined as:

$$z_p^0 = [x - w_1 \ y - w_2 \ z - w_3]^T \tag{14.55}$$

then its first derivative can be defined as:

$$\dot{z}_p^0 = z_p^1 = [\dot{x} - \dot{w}_1 \ \dot{y} - \dot{w}_2 \ \dot{z} - \dot{w}_3]^T \tag{14.56}$$

where the expressions of $\dot{x}$, $\dot{y}$ and $\dot{z}$ are given in Equation (14.18). Since the real disturbances are replaced by their estimations in the closed-loop system, we define the next state as:

$$z_p^2 = \mathbf{R}_b^i \begin{bmatrix} \hat{d}_x \\ \hat{d}_y \\ T + \hat{d}_z \end{bmatrix} + \begin{bmatrix} 0 \\ 0 \\ g \end{bmatrix} - \begin{bmatrix} \ddot{w}_1 \\ \ddot{w}_2 \\ \ddot{w}_3 \end{bmatrix}, \tag{14.57}$$

Invoking Equation (14.18) and (14.20) gives $\dot{z}_p^1 = z_p^2 + \mathbf{R}_b^i \cdot e_{d_f}$. Similarly, z_p^3 is defined as:

$$z_p^3 = \mathbf{R}_b^i \hat{\omega} \begin{bmatrix} \hat{d}_x \\ \hat{d}_y \\ T + \hat{d}_z \end{bmatrix} + \mathbf{R}_b^i \begin{bmatrix} 0 \\ 0 \\ Z_w \dot{w} + Z_{col} \dot{\delta}_{col} \end{bmatrix} - \begin{bmatrix} \dddot{w}_1 \\ \dddot{w}_2 \\ \dddot{w}_3 \end{bmatrix} \tag{14.58}$$

From Equation (14.57) and (14.58) and recalling observer dynamics (14.49), it can be observed that

$$\dot{z}_p^2 = z_p^3 - \mathbf{R}_b^i \dot{e}_{d_f} = z_p^3 + \mathbf{R}_b^i L_1 e_{d_f} \tag{14.59}$$

Repeat this procedure, z_p^4 is defined from Equation (14.24) by using estimated disturbances, such that

$$\dot{z}_p^3 = z_p^4 + \mathbf{R}_b^i \hat{\omega} \cdot L_1 e_{d_f} \tag{14.60}$$

In addition, the heading tracking error and its derivatives are defined as $z_\psi^0 = \psi - w_4$, $z_\psi^1 = \dot{\psi} - \dot{w}_4$, where $\dot{\psi}$ is given in Equation (14.19), and $z_\psi^2 = \ddot{\psi} - \ddot{w}_4$, where $\ddot{\psi}$ is provided in Equation (14.21).

Finally, by invoking Equation (14.26) and the definitions of z_p^4 and z_ψ^2, we have

$$
\begin{aligned}
\begin{bmatrix} z_p^4 \\ z_\psi^2 \end{bmatrix} &= \hat{M}_1 + A(\mathbf{x}, \hat{\mathbf{d}}_f)(\mathbf{u} + \mathbf{u}_0) \\
&= \hat{M}_1 + A(\mathbf{x}, \hat{\mathbf{d}}_f)(-A(\mathbf{x}, \hat{\mathbf{d}}_f)^{-1}(K\hat{M}_\rho + \hat{M}_1) - \hat{\mathbf{u}}_0 + \mathbf{u}_0) \\
&= -K\hat{M}_\rho + A(\mathbf{x}, \hat{\mathbf{d}}_f)e_{u_0}
\end{aligned} \tag{14.61}
$$

where, $e_{u_0} = \mathbf{u}_0 - \hat{\mathbf{u}}_0$ and K has the form:

$$
K = \begin{bmatrix}
k_{11} \dots k_{14} & 0_{1\times4} & 0_{1\times4} & 0_{1\times2} \\
0_{1\times4} & k_{21} \dots k_{24} & 0_{1\times4} & 0_{1\times2} \\
0_{1\times4} & 0_{1\times4} & k_{31} \dots k_{34} & 0_{1\times2} \\
0_{1\times4} & 0_{1\times4} & 0_{1\times4} & k_{41} \dots k_{42}
\end{bmatrix} \tag{14.62}
$$

By recalling the definition of $\hat{M}_\rho$ in Equation (14.45), Equation (14.61) can be written as:

$$
\begin{bmatrix} \dot{z}_p^3 \\ \dot{z}_\psi^1 \end{bmatrix} = \begin{bmatrix} K_1 z_p^0 + K_2 z_p^1 + K_3 z_p^2 + K_4 z_p^3 \\ k_{41} z_\psi^0 + k_{42} z_\psi^1 \end{bmatrix} + \begin{bmatrix} \mathbf{R}_b^i \hat{\omega} \cdot L_1 e_{d_f} \\ 0 \end{bmatrix} + A(\mathbf{x}, \hat{\mathbf{d}}_f)e_{u_0} \tag{14.63}
$$

where $K_i = \mathrm{diag}(k_{1i}, k_{2i}, k_{3i})$, for $i = 1, 2, 3, 4$.

Summarizing Equation (14.55) Equation (14.63) gives a linear form of the closed-loop system:

$$
\begin{bmatrix} \dot{z}_p^0 \\ \dot{z}_p^1 \\ \dot{z}_p^2 \\ \dot{z}_p^3 \\ \dot{z}_\psi^0 \\ \dot{z}_\psi^1 \end{bmatrix} = \underbrace{\begin{bmatrix} 0_{3\times3} & I_3 & 0_{3\times3} & 0_{3\times3} & 0 & 0 \\ 0_{3\times3} & 0_{3\times3} & I_3 & 0_{3\times3} & 0 & 0 \\ 0_{3\times3} & 0_{3\times3} & 0_{3\times3} & I_3 & 0 & 0 \\ K_1 & K_2 & K_3 & K_4 & 0 & 0 \\ 0_{1\times3} & 0_{1\times3} & 0_{1\times3} & 0_{1\times3} & 1 & 0 \\ 0_{1\times3} & 0_{1\times3} & 0_{1\times3} & 0_{1\times3} & k_{41} & k_{41} \end{bmatrix}}_{A_z} \underbrace{\begin{bmatrix} z_p^0 \\ z_p^1 \\ z_p^2 \\ z_p^3 \\ z_\psi^0 \\ z_\psi^1 \end{bmatrix}}_{z} + \underbrace{\begin{bmatrix} 0_{3\times1} \\ \epsilon_1 \\ \epsilon_2 \\ \epsilon_3 \\ 0_{1\times1} \\ \epsilon_5 \end{bmatrix}}_{\epsilon} \qquad (14.64)
$$

or, compactly

$$
\dot{z} = A_z z + \epsilon \qquad (14.65)
$$

where $\epsilon_1 = R_b^i \cdot e_{d_f}$, $\epsilon_2 = R_b^i \cdot L_1 e_{d_f}$, $\epsilon_3 = R_b^i \hat{\omega} \cdot L_1 e_{d_f} + \overline{A}(x, \hat{d}_f) e_{u_0}$ and $\epsilon_5 = \underline{A}(x, \hat{d}_f) e_{u_0}$. All these terms depend on the helicopter states and estimation errors e_d.

System (14.64) can be classified as a cascade system in the new coordinates:

$$
\begin{aligned}
\dot{z} &= f_1(z) + \epsilon(x, e_d) e_d \\
\dot{e}_d &= f_2(e_d)
\end{aligned} \qquad (14.66)
$$

where the upper system is Equation (14.64) and the lower system is the observer dynamics (14.49).

When estimation errors are zeros, the upper system $\dot{z} = f_1(z)$ reduces to a linear system $\dot{z} = A_z z$. Its global asymptotic stability can be guaranteed by proper choosing MPC gain K such that A_z is Hurwitz. In this case, it can be achieved by setting control order $r > 2$ [40]. On the other hand, the global asymptotic stability of the lower system is guaranteed during the design of disturbance observer by letting $L > 0$. Therefore, the closed-loop system under the composite control law is at least locally asymptotically stable according to [106]. We can extend the above result further by introducing the following lemma.

Assumption 14.1 *The system $\dot{z} = f_1(z)$ is globally uniformly asymptotically stable with a Lyapunov function $V(z)$, $V : \mathbb{R}^n \to \mathbb{R}$ positive definite (that is $V(0) = 0$ and $V(z) > 0$ for all $z \neq 0$) and proper which satisfies*

$$
\left\| \frac{\partial V}{\partial z} \right\| \|z\| \leq c_1 V(z), \; \forall \, \|z\| \geq \eta \qquad (14.67)
$$

where $c_1 > 0$ and $\eta > 0$.

Assumption 14.2 *The function $\epsilon(x, e_d)$ satisfies*

$$\|\epsilon(x, e_d)\| \leq \alpha_1(\|e_d\|) + \alpha_2(\|e_d\|) \|z\| \tag{14.68}$$

where $\alpha_1, \alpha_2 : \mathbb{R} \rightarrow \mathbb{R}$ are continuous.

Assumption 14.3 *Equation $\dot{e}_d = f_2(e_d)$ is globally uniformly asymptotically stable and for all $t \geq t_0$*

$$\int_{t_0}^{\infty} \|e_d(t)\| \, dt \leq \beta(\|e_d(t_0)\|) \tag{14.69}$$

where function β is a class $\mathcal{K}$ function.

Lemma 14.1 *[198] If Assumptions 14.1–14.3 below are satisfied then the cascaded system (14.66) is globally uniformly asymptotically stable.*

The rigorous proof of Lemma 14.1 is given in [198]. The basic idea is first to show that the upper system of cascade system does not escape to infinite in finite time and are bounded for $t > t_0$ with the condition that the input vector $\epsilon(x, e_d)$ grows linearly and at the fastest in the state z. Then it needs to show that as $t \rightarrow \infty$, estimation error $e_d \rightarrow 0$ and $z \rightarrow 0$ due to the global asymptotic stability of $\dot{z} = f_1(z)$.

Theorem 14.1 *Given that the reference trajectory $\mathbf{w}$, its first ρ_i derivatives, and disturbances $\mathbf{d}$ are bounded, the closed-loop system (14.61) under the composite control is globally asymptotically stable.*

Proof By using Lemma 14.1, for closed-loop system (14.61) in the cascade form (14.66) if Assumptions 14.1–14.3 are satisfied, the proof will then be completed.

First, Assumption 14.1 is satisfied due to the fact that $\dot{z} = f_1(z) = A_z z$ and A_z is Hurwitz. Then, we investigate the boundness on $\epsilon(x, e_d)$ in terms of $\|z\|$ and $\|e_d\|$. From their definitions, we have

$$\|\epsilon_1\| \leq \|e_d\| \tag{14.70}$$

$$\|\epsilon_2\| \leq \|L_1\| \|e_d\| \tag{14.71}$$

$$\|\epsilon_3\| \leq \|\hat{\omega}\| \|L_1\| \|e_{d_f}\| + \|\overline{A}(\cdot, \cdot)\| \|e_d\| \tag{14.72}$$

$$\|\epsilon_5\| \leq \|\underline{A}(\cdot, \cdot)\| \|e_d\| \tag{14.73}$$

The skew-matrix $\hat{\omega}$ can be seen consisted of nominal state decided by the reference command and the error state, i.e., $\hat{\omega} = \hat{\omega}_c + \hat{\omega}_e$. The former one is bounded as the bounded command, and the latter one is bounded by tracking error $\|z\|$. Therefore, there exist two constants, $b_1 > 0$ and $b_2 > 0$, such that $\|\dot{\omega}\| \leq b_1 + b_2 \|z\|$. Moreover, $\|\overline{A}(\cdot, \cdot)\|$ linearly depends on $\dot{d}$ and state T. Due to d is bounded and disturbance observer is globally exponentially stable, $\dot{d}$ is also bounded. On the other hand, T is bounded according to Equation (14.57). Hence, we have $\|\overline{A}(\cdot, \cdot)\| \leq b_3 + b_4 \|z\|$, for some $b_3 > 0$ and $b_4 > 0$. Then, the bound on ϵ_3 can be written as $\|\epsilon_3\| \leq \beta_1 \|e_d\| + \beta_2 \|e_d\| \|z\|$, for some $\beta_1 > 0$ and $\beta_2 > 0$. At last, following the same fashion, $\epsilon_5 \leq b_5 \|e_d\|$, for some $b_5 > 0$ if pitch angle $\theta \neq \pm\pi/2$. Combining bounds on $\|\epsilon_i\|$, $i = 1, \ldots, 5$ gives

$$
\begin{aligned}
\|\epsilon\| &\leq \|\epsilon_1\| + \ldots + \|\epsilon_5\| \\
&\leq \gamma_1 \|e_d\| + \gamma_2 \|e_d\| \|z\|
\end{aligned}
\tag{14.74}
$$

where $\gamma_1 > 0$ and $\gamma_2 > 0$. Thus, Assumption 14.2 is satisfied.

Finally, as lower system $\dot{e}_d = f_2(e_d)$ is globally exponentially stable, Assumption 14.3 is satisfied. And the proof of this theorem is completed.

14.6 Simulation and Experiment

The simulation and experiment are based on a Trex-250 miniature helicopter, which is a 200-sized helicopter with a main rotor diameter of 460 mm and a trail rotor diameter of 108 mm. The miniaturized size and 3D aerobatic ability make it well-suited for indoor flight tests. Moreover, Trex-250 has a collective pitch rotor and well-designed Bell-Hiller stabilizer mechanism that makes it representative of most widely used small-scale helicopters.

Numerical simulations are first carried out to investigate the proposed control framework. In the simulation, it is assumed that there are 20% uncertainties on the model parameters. Furthermore, there is a constant wind disturbance with a speed of $5m/s$ acting on the helicopter. The helicopter is required to track an eight-shape trajectory with and without the compensation of DOBC. The tracking results are shown in Figure 14.4 and the corresponding cyclic controls are given in Figure 14.5.

It can be seen from the simulation that the ENMPC is able to deal with uncertainties and achieve satisfactory tracking, but it cannot compensate the steady-state error mainly caused by the wind disturbance. In contrast, the action of DOBC, taking into account the disturbances from both external and internal sources, eliminates the tracking error and smooths the control signals.

Several flight tests have been designed to investigate the control performance of the proposed scheme on real helicopter. The first test presented here is a hovering and perturbation test. The helicopter was required to take-off and hover at the origin at height of 0.5m. A wind perturbation was then applied on the helicopter by posing a fan in front of the helicopter (see Figure 14.6).

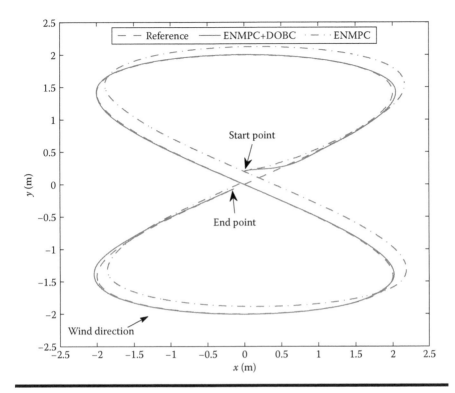

Figure 14.4 Eight-shape tracking.

The test result is given in Figure 14.7. In the test, the helicopter was first under the control of ENMPC to perform take-off and hovering. It can be seen that the ENMPC stabilized the helicopter but with a steady-state error due to the mismatch between the model used for ENMPC design and the real helicopter dynamics. After 25 seconds, the disturbance observer switched on and the composite controller took action to bring the helicopter to the setpoint. After 60 seconds, the fan was turned on to generate the wind gust. The average wind speed is about 3m/s, which is significant strong for our test helicopter with a small dimension. This can be observed from the attitude history in Figure 14.10, where the magnitude of pitch and roll angles of the helicopter dramatically surged after wind gust is applied. However, the composite controller exhibited an excellent performance under the wind gust and maintained the helicopter position very well. The force disturbances estimated by are given in Figure 14.8, and the control signals are illustrated in Figure 14.9.

It is also interesting to see where the disturbances come from without external wind gust, and this will also explain why ENMPC based on the nominal model cannot achieve asymptotic tracking if the helicopter is not trimmed properly. By recalling the helicopter dynamics model (14.1) and considering the steady-state

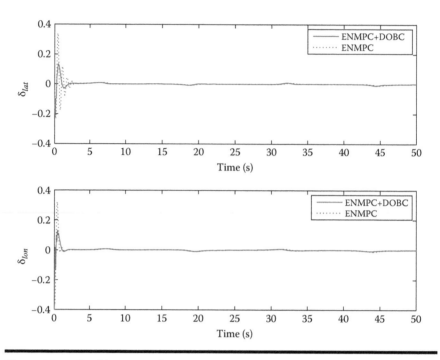

Figure 14.5   Tracking controls.

Figure 14.6   Hovering and perturbation test.

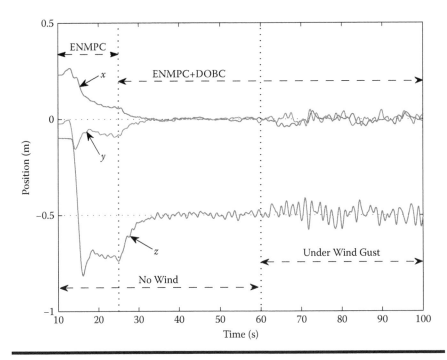

Figure 14.7   Helicopter position.

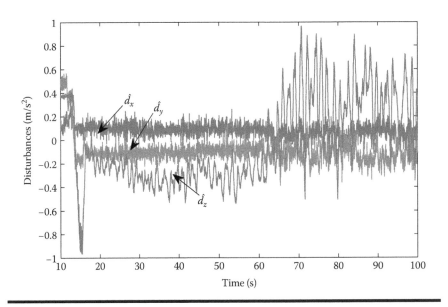

Figure 14.8   Disturbance observer outputs.

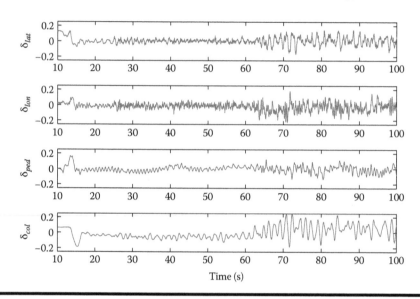

Figure 14.9   Control signals.

model, we have

$$0 = -g \sin \theta_0 + d_x$$
$$0 = g \cos \theta_0 \sin \phi_0 + d_y \tag{14.75}$$

where ϕ_0 and θ_0 are the trim attitude depending on the particular helicopter. The trim attitude may be attributed to asymmetrical structure and model uncertainties. Their values are small and usually can be ignored in the theoretical analysis, but they do affect the actual control performance as they project vertical lift to longitudinal and lateral directions. This phenomena can be further explained by carefully examining the measurement from the flight test. Observing the attitude measurement in Figure 14.10 reveals the average roll and pitch angles are about 0.01 rad, which contribute $0.1 \, m/s^2$ and $-0.1 \, m/s^2$ to d_x and d_y, according to Equation (14.75). The estimated $\hat{d}_x$ and $\hat{d}_y$ from the observer are very close to our rough calculation. This quantitative analysis gives good confidence for the proposed disturbance observer.

Unlike the conventional MPC being restricted to a low-control bandwidth, the high bandwidth that ENMPC can reach makes it a suitable candidate for controlling a helicopter to perform acrobatic manoeuvres. In the second flight test, the helicopter was controlled to perform a pirouette manoeuvre, in which a helicopter started from the hovering position and flew along a straight line while pirouetting at a yaw rate of 120 deg/sec. This is an extremely challenging flight pattern, because the lateral and longitudinal controls are strongly coupled by the rotation, and they have to coordinate with each other to achieve a straight progress. Besides, the varying position

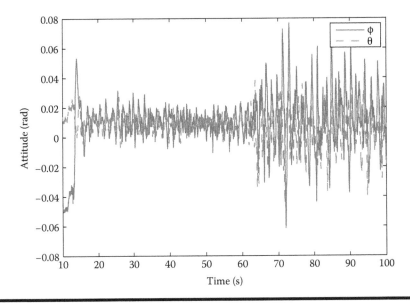

Figure 14.10   Helicopter attitude.

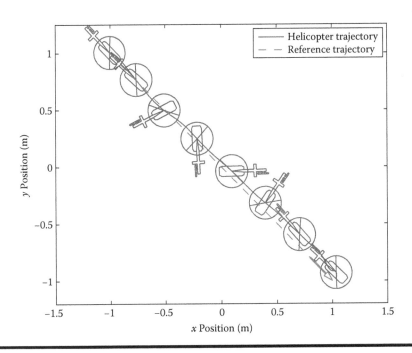

Figure 14.11   Pirouette manoeuvre results.

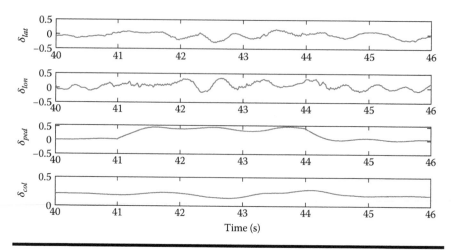

Figure 14.12   Control signals.

of the tail rotor with respect to the progress direction poses severe disturbances on the forward flight. The result from the flight test is shown in Figure 14.11 and control signals are provided in Figure 14.12. It can be seen that the helicopter under the control of ENMPC executed the manoeuvre with high quality.

14.7 Summary

This chapter has described a composite control framework for trajectory tracking of autonomous helicopters. The nonlinear tracking control has been achieved by an explicit MPC algorithm, which has eliminated the computational-intensive online optimisation in the traditional MPC. On the implementation side, the introducing of disturbance observer has solved the difficulties of applying model-based control technique into the practical environment. The design of ENMPC has provided a seamless way of integrating the disturbance information. On the other hand, the robustness and disturbance attenuation of the controller has been enhanced by the nonlinear disturbance observer.

Simulation and experiment results show promising performance of the combination of ENMPC and DO. Apart from the reliable tracking that the proposed controller guarantees, it also has the ability of estimating the helicopter trim condition during the flight, which helps the controller to deal with the variation of the helicopter status like payload changing and component upgrades.

Chapter 15

Disturbance Rejection for Bank-to-Turn Missiles

15.1 Introduction

Due to the growing interest in missiles with long-range high maneuverability and precision, the bank-to-turn (BTT) steering technique has become more and more popular as compared with the traditional skid-to-turn (STT) method [199, 52]. By orienting the maximum aerodynamic normal force to the desired direction rapidly with a substantially large roll rate, the BTT missile exhibits many advantages over the STT missile, including high-lift, low-drag, air-intake, internal carriage, and increased range [199, 42, 53, 43]. However, the structure configuration of the BTT missile has undergone significant changes to take its advantages. These changes bring many challenges to autopilot designers, e.g., (1) the effects caused by time-varying parameters are more severe as compared with the STT style, (2) the high roll rate and the asymmetric structure will inevitably induce heavy nonlinear crossing couplings between different channels, and (3) the input/output dynamic characteristics for certain channels are nonminimum phase [199, 200].

To this end, autopilot design for BTT missiles has attracted extensive attention, and various strategies have been proposed. Based on the models obtained by approximate linearization in given flight conditions or input/output feedback linearization technique, a few elegant control methods including robust control [42], gain-scheduling control [45, 46], nonlinear control [199], model predictive control [49], and switching control [201] have been employed for advanced autopilot designs. Performance, stability, and robustness of modern missiles have been significantly improved with these advanced control techniques. Further development is still required to meet even increasing performance requirements for advanced missiles.

Feedforward control is known as an active approach in rejecting disturbances if the disturbance is measurable [20]. However, the uncertainties and disturbances in the missile systems are generally complicated and unmeasurable, thus an effective solution is to develop disturbance estimation techniques. This chapter investigates a disturbance observer-based control (DOBC) approach to improve the disturbance-rejection ability of the BTT missiles.

By estimating and compensating the effects caused by disturbances/uncertainties, DOBC provides an active approach to handle system disturbances and improve robustness [78, 3, 86, 2] in the presence of unmeasurable uncertainties. In the past few decades, DOBC approaches have been successfully researched and applied in various engineering fields including robotic systems [2, 202], position systems [29, 203], grinding systems [20, 1], and flight control systems [3, 86, 204, 93]. Compared with other robust control schemes, DOBC approach has two distinct features. One feature is that disturbance observer-based compensation can be considered as a "patch" for existing controllers that may have unsatisfactory disturbance attenuation and robustness against uncertainties. The benefits of this are that there is no change to a baseline controller that may have been widely used and developed for many years. The second feature is that DOBC is not a worst-case-based design. Most of the existing robust control methods are worst-case-based design, and have been criticized as being "over conservative." In the DOBC approach, the nominal performance of the baseline controller is recovered in the absence of disturbances or uncertainties.

A disturbance observer-based robust control (DOBRC) method is proposed to solve the disturbance attenuation problem of the BTT missile system. In the BTT missiles, the lumped disturbance torques caused by unmodeled dynamics, external winds, and parameter variations may affect the states directly rather than through the input channels, therefore don't satisfy matching conditions [100]. Using the DOBRC method, autopilot design of the BTT missile is carried out in this chapter. The missile system under consideration is subject to mismatched disturbances including not only external disturbances but also model uncertainties.

15.2 Pitch/Yaw Dynamic Models of BTT Missiles

Since the roll rate in the BTT missile is much larger than the pitch and yaw rates, the pitch and yaw dynamics are severely interrupted by the roll dynamic. While the pitch and yaw dynamics have relatively smaller effects on the roll channel. To this end, a widely used method in the BTT missile is to design an autopilot for the pitch and yaw channels together, and design an autopilot for the roll channel separately [201, 47]. In this chapter, only an autopilot design for pitch/yaw channels are considered as the roll channel design is relatively easy. The pitch/yaw dynamic models of the BTT missile are taken from [201, 47], depicted by

$$\begin{cases} \dot{\omega}_z = -(a_1 + e_1)\omega_z + (e_1 a_4 - a_2)\alpha + \frac{e_1}{57.3}\omega_x\beta + (-e_1 a_5 - a_3)\delta_z \\ \qquad + \frac{J_x - J_y}{57.3 J_z}\omega_x\omega_y + d_{\omega_z}, \\ \dot{\alpha} = \omega_z - \frac{1}{57.3}\omega_x\beta - a_4\alpha - a_5\delta_z + d_\alpha, \\ \dot{\omega}_y = -(b_1 + e_2)\omega_y + (e_2 b_4 - b_2)\beta - \frac{e_2}{57.3}\omega_x\alpha + (e_2 b_5 - b_3)\delta_y \\ \qquad + \frac{J_z - J_x}{57.3 J_y}\omega_x\omega_z + d_{\omega_y}, \\ \dot{\beta} = \omega_y + \frac{1}{57.3}\omega_x\alpha - b_4\beta - b_5\delta_y + d_\beta, \\ n_z = -\frac{b_4 v_t}{57.3 g}\beta - \frac{b_5 v_t}{57.3 g}\delta_y + d_{n_z}, \\ n_y = \frac{a_4 v_t}{57.3 g}\alpha + \frac{a_5 v_t}{57.3 g}\delta_z + d_{n_y}, \end{cases} \qquad (15.1)$$

where ω_x, ω_y and ω_z are roll, yaw, and pitch rates, respectively. Variables α and β represent the angle-of-attack and the sideslip angle, while δ_y and δ_z are yaw and pitch control deflection angles, respectively; n_z and n_y denote the overloads on the normal and side direction. d_{ω_z}, d_α, d_{ω_y}, d_β, d_{n_z}, d_{n_y} denote external disturbances on each equation. Parameters J_x, J_y, and J_z denote roll, yaw, and pitch moments of inertia, respectively. v_t and g are the instantaneous speed and the gravity acceleration. Coefficients a_i, $b_i (i = 1, 2, \dots, 5)$, e_1, and e_2 are aerodynamic parameters of the missile systems.

During the flight process of the missile, the aerodynamic parameters vary with the change of the missile height and velocity. In addition, the parameter perturbations are very complex, and it is almost impossible to obtain their analytic forms. The aerodynamic parameters for different operating points are taken from [201] and listed in Table 15.1.

Since the roll rate is much larger than the pitch and yaw rates, it is reasonable to take ω_x as a parameter of the pitch/yaw models [47] and the following linear model is obtained by reformulating Equation (15.1) as

$$\begin{cases} \dot{x} = Ax + B_u u + B_d d_x, \\ y = Cx + D_u u + D_d d_y, \end{cases} \qquad (15.2)$$

with matrices

$$A = \begin{bmatrix} -a_1 - e_1 & e_1 a_4 - a_2 & \frac{(J_x - J_y)\omega_x}{57.3 J_z} & \frac{e_1 \omega_x}{57.3} \\ 1 & -a_4 & 0 & \frac{-\omega_x}{57.3} \\ \frac{(J_z - J_x)\omega_x}{57.3 J_y} & \frac{-e_2 \omega_x}{57.3} & -b_1 - e_2 & e_2 b_4 - b_2 \\ 0 & \frac{\omega_x}{57.3} & 1 & -b_4 \end{bmatrix}, \qquad (15.3)$$

$$B_u = \begin{bmatrix} -e_1 a_5 - a_3 & 0 \\ -a_5 & 0 \\ 0 & e_2 b_5 - b_3 \\ 0 & -b_5 \end{bmatrix}, \qquad (15.4)$$

Table 15.1 Aerodynamic Parameters of a BTT Missile for Different Operating Points

Operating points	$t_1(4.4s)$	$t_2(11.7s)$	$t_3(19.5s)$	$t_4(23s)$
a_1	1.593	1.485	1.269	1.130
a_2	260.559	266.415	196.737	137.385
a_3	185.488	182.532	176.932	160.894
a_4	1.506	1.295	1.169	1.130
a_5	0.298	0.243	0.217	0.191
b_1	1.655	1.502	1.269	1.130
b_2	39.988	−24.627	−31.452	−41.425
b_3	159.974	170.532	182.030	184.093
b_4	0.771	0.652	0.680	0.691
b_5	0.254	0.191	0.188	0.182
e_1	0.285	0.192	0.147	0.118
e_2	0.295	0.195	0.147	0.118

$$\mathbf{C} = \begin{bmatrix} 0 & 0 & 0 & \frac{-b_4 v_t}{57.3g} \\ 0 & \frac{a_4 v_t}{57.3g} & 0 & 0 \end{bmatrix}, \tag{15.5}$$

$$\mathbf{D}_u = \begin{bmatrix} 0 & \frac{-b_5 v_t}{57.3g} \\ \frac{a_5 v_t}{57.3g} & 0 \end{bmatrix}, \tag{15.6}$$

$\mathbf{B}_d = \mathbf{I}_{4\times4}$, $\mathbf{D}_d = \mathbf{I}_{2\times2}$, the state vector $\boldsymbol{x} = [\omega_z\ \alpha\ \omega_y\ \beta]^T$, the control input $\boldsymbol{u} = [\delta_z\ \delta_y]^T$, the output $\boldsymbol{y} = [n_z\ n_y]^T$, the external disturbances on states $\boldsymbol{d}_x = [d_{\omega_z}\ d_\alpha\ d_{\omega_y}\ d_\beta]^T$, and outputs $\boldsymbol{d}_y = [d_{n_z}\ d_{n_y}]^T$.

The autopilot design objective is to achieve a good overload tracking performance according to the guidance commands as well as robustness against external disturbances and model uncertainties caused by parameter variations. In addition, from the viewpoint of engineering practice, the sideslip angle should be restricted as $|\beta| < 5$ deg, and the actuator deflection angles should be limited within a region $|\delta_{max}| \leq 21$ deg [41].

15.3 Disturbance Observers

In the framework of the disturbance observer technique, not only external disturbances but also the influence caused by model uncertainties can be estimated and attenuated. To facilitate this, the model (15.2) is rewritten as

$$\begin{cases} \dot{x} = A_n x + B_n u + B_{ld} d_{lx}, \\ y = C_n x + D_n u + D_{ld} d_{ly}, \end{cases} \tag{15.7}$$

where A_n, B_n, C_n, and D_n are system matrices in the nominal case and d_{lx} and d_{ly} denote the lumped disturbances on states and outputs. Letting $B_{ld} = I_{4\times4}$ and $D_{ld} = I_{2\times2}$ and comparing (15.2) with (15.7) yields

$$d_{lx} = B_d d_x + (A - A_n)x + (B - B_n)u, \tag{15.8}$$

$$d_{ly} = D_d d_y + (C - C_n)x + (D - D_n)u. \tag{15.9}$$

The lumped disturbances consist of external disturbances and internal disturbances caused by model uncertainties. Disturbance observers are now designed to estimate the disturbances d_{lx} and d_{ly} using the input, output, and state information. The estimate of d_{ly} can be obtained from the second equation in (15.7), described by

$$\hat{d}_{ly} = y - C_n x - D_n u. \tag{15.10}$$

Combining (15.7) with (15.10), the estimation error $e_{d_{ly}}$ can be obtained as

$$e_{d_{ly}} = \hat{d}_{ly} - d_{ly} = 0. \tag{15.11}$$

Assumption 15.1 *Suppose that the lumped disturbances d_{lx} varies slowly relative to the observer dynamics and has constant steady-state values* $\lim_{t\to\infty} \dot{d}_{lx}(t) = 0$ *or* $\lim_{t\to\infty} d_{lx}(t) = d_{lxs}$.

Remark 15.1 *The results in this chapter are based on Assumption 15.1. However, it is shown that the method is also feasible for fast time-varying disturbances [2].*

Remark 15.2 *In the presence of uncertainties, the lumped disturbances would be a function of the states, which can be reasonably estimated if the disturbance observer dynamics are faster than that of the closed-loop system. The same argument for the state observer based control methods is valid.*

For system (15.7), the following observer is designed to estimate the disturbances d_{lx}

$$\begin{cases} \dot{z} = -L(z + Lx) - L(A_n x + B_n u), \\ \hat{d}_{lx} = z + Lx, \end{cases} \tag{15.12}$$

where $\hat{\boldsymbol{d}}_{lx}$ is the estimate of the lumped disturbance on the states $\boldsymbol{d}_{lx}$, $\boldsymbol{z}$ is an auxiliary vector, and $\mathbf{L}$ is the observer gain matrix to be designed.

Lemma 15.1 *[12]: Consider system (15.7) under the lumped disturbances that satisfy Assumption 15.1. The estimates of the disturbance observer (15.12) asymptotically tracks the lumped disturbances if the observer gain matrix $\mathbf{L}$ is chosen such that $-\mathbf{L}$ is a Hurwitz matrix.*

Proof The disturbance estimation error of the DOB (15.12) is defined as

$$\boldsymbol{e}_{d_{lx}} = \hat{\boldsymbol{d}}_{lx} - \boldsymbol{d}_{lx}. \tag{15.13}$$

Combining (15.7), (15.12) with (15.13) gives

$$
\begin{aligned}
\dot{\boldsymbol{e}}_{d_{lx}} &= \dot{\hat{\boldsymbol{d}}}_{lx} - \dot{\boldsymbol{d}}_{lx} \\
&= \dot{\boldsymbol{z}} + \mathbf{L}\dot{\boldsymbol{x}} - \dot{\boldsymbol{d}}_{lx} \\
&= -\mathbf{L}\hat{\boldsymbol{d}}_{lx} - \mathbf{L}(\mathbf{A}_n\boldsymbol{x} + \mathbf{B}_n\boldsymbol{u}) \\
&\quad + \mathbf{L}(\mathbf{A}_n\boldsymbol{x} + \mathbf{B}_n\boldsymbol{u} + \boldsymbol{d}_{lx}) - \dot{\boldsymbol{d}}_{lx} \\
&= -\mathbf{L}(\hat{\boldsymbol{d}}_{lx} - \boldsymbol{d}_{lx}) - \dot{\boldsymbol{d}}_{lx} = -\mathbf{L}\boldsymbol{e}_{d_{lx}} - \dot{\boldsymbol{d}}_{lx}.
\end{aligned}
\tag{15.14}
$$

Since all eigenvalues of matrix $-\mathbf{L}$ are in the left hand side of the complex plane, Equation (15.14) is asymptotically stable. This implies that the estimate of DOB can track the disturbances asymptotically under the condition that $\lim_{t\to\infty} \dot{\boldsymbol{d}}_{lx}(t) = \boldsymbol{0}_{4\times1}$.
□

Since $\mathbf{B}_n \neq \mathbf{I}_{4\times4}$, $\mathbf{D}_n \neq \mathbf{I}_{2\times2}$, it can be found from Equation (15.7) that the disturbances and uncertainties are "mismatched" ones. The existing DOBC methods are not applicable in such a case.

It should be pointed out that the "mismatched" disturbances can not be attenuated from the state equations generally. In this chapter, based on the disturbance estimate of DOB, the composite control law like $\boldsymbol{u} = \mathbf{K}_x\boldsymbol{x} + \mathbf{K}_d\hat{\boldsymbol{d}}$ is designed, and we attempt to find an appropriate $\mathbf{K}_d$ to assure that the disturbances can be removed from the output channel in steady-state.

15.4 Disturbance Observer-Based Robust Control

15.4.1 Feedback Control Design

In the proposed composite DOBRC method, any feedback controller that can stabilize system (15.7) and provide adequate overload tracking performance in the absence of disturbances and uncertainties can be adopted. The classical linear quadratic regulator (LQR) is considered as the baseline autopilot design for the BTT missiles.

The penalty matrix $\mathbf{Q}$ and $\mathbf{R}$ in the cost function of LQR are selected as

$$\mathbf{Q} = \begin{bmatrix} 0.0882 & 0.0128 & 0.0264 & 0.02 \\ 0.0128 & 0.204 & 0.004 & 0.0088 \\ 0.0264 & 0.004 & 0.018 & 0.0064 \\ 0.02 & 0.0088 & 0.0064 & 1.5208 \end{bmatrix}, \mathbf{R} = \begin{bmatrix} 2 & 0 \\ 0 & 30 \end{bmatrix}. \tag{15.15}$$

The corresponding LQR feedback gain is obtained as

$$\mathbf{K}_x = \begin{bmatrix} 0.2007 & -0.4539 & 0.0808 & 0.1606 \\ 0.0047 & -0.0436 & 0.0267 & 0.1306 \end{bmatrix}. \tag{15.16}$$

As shown in Figures 15.2–15.4, a satisfactory step response is obtained (at the beginning of the responses).

15.4.2 Stability Analysis of Closed-Loop System

Different from all previous DOBC methods, our new DOBRC law for system (15.7) is designed as

$$u = \mathbf{K}_x x + \mathbf{K}_d \hat{d}, \tag{15.17}$$

where $\mathbf{K}_d = [\mathbf{K}_{dx} \ \mathbf{K}_{dy}]$, and $\hat{d} = [\hat{d}_{lx} \ \hat{d}_{ly}]^T$.

Assumption 15.2 *The lumped disturbances satisfy that $d_{lx}, d_{ly},$ and $\dot{d}_{lx}$ are bounded.*

Lemma 15.2 *[12]: Suppose that Assumption 15.2 is satisfied. The closed-loop system consisting of the plant (15.7), the DO (15.12), and the composite control law (15.17) is bounded-input-bounded-output (BIBO) stable if the feedback control gain $\mathbf{K}_x$, and the observer gain matrix $\mathbf{L}$ are selected such that (1) the system (15.7) in the absence of disturbances under control law $u = \mathbf{K}_x x$ is asymptotically stable, and (2) the disturbance estimate by DOB can track the disturbance asymptotically.*

Proof Combining system (15.7) with the DOBRC law (15.17) and disturbance observers (15.10)–(15.13), the closed-loop system is obtained as

$$\begin{bmatrix} \dot{x} \\ \dot{e}_{d_{lx}} \\ \dot{e}_{d_{ly}} \end{bmatrix} = \begin{bmatrix} \mathbf{A}_n + \mathbf{B}_n \mathbf{K}_x & \mathbf{B}_n \mathbf{K}_{dx} & \mathbf{B}_n \mathbf{K}_{dy} \\ \mathbf{O} & -\mathbf{L} & \mathbf{O} \\ \mathbf{O} & \mathbf{O} & \mathbf{O} \end{bmatrix} \begin{bmatrix} x \\ e_{d_{lx}} \\ e_{d_{ly}} \end{bmatrix}$$
$$+ \begin{bmatrix} \mathbf{I}_{4\times4} + \mathbf{B}_n \mathbf{K}_{dx} & \mathbf{B}_n \mathbf{K}_{dy} & \mathbf{O} \\ \mathbf{O} & \mathbf{O} & -\mathbf{I}_{4\times4} \\ \mathbf{O} & \mathbf{O} & \mathbf{O} \end{bmatrix} \begin{bmatrix} d_{lx} \\ d_{ly} \\ \dot{d}_{lx} \end{bmatrix}. \tag{15.18}$$

Considering the condition $e_{d_{ly}} = 0$ for all $t \geq 0$ in (15.11), the closed-loop system in (15.18) reduces to

$$\begin{bmatrix} \dot{x} \\ \dot{e}_{d_{lx}} \end{bmatrix} = \begin{bmatrix} A_n + B_n K_x & B_n K_{dx} \\ O & -L \end{bmatrix} \begin{bmatrix} x \\ e_{d_{lx}} \end{bmatrix} + \begin{bmatrix} I_{4\times4} + B_n K_{dx} & O \\ O & -I_{4\times4} \end{bmatrix} \begin{bmatrix} d_{lx} \\ \dot{d}_{lx} \end{bmatrix}.$$

(15.19)

Conditions (1) and (2) imply that both $A_n + B_n K_x$ and $-L$ are Hurwitz matrices. To this end, it can be proved that

$$\begin{bmatrix} A_n + B_n K_x & B_n K_{dx} \\ O & -L \end{bmatrix}$$

is also a Hurwitz matrix.

Moreover, Assumption 15.2 implies that the inputs of the closed-loop system (15.19) d_{lx} and $\dot{d}_{lx}$ are bounded. This completes the proof that the outputs of the closed-loop system are bounded.

Lemma 15.3 *[12]: Suppose that both Assumption 15.1 and 15.2 are satisfied. The state vector x converges to a constant vector $x_s = -(A_n + B_n K_x)^{-1}(I_{4\times4} + B_n K_x)d_{lxs}$ asymptotically if the feedback control gain K_x and the observer gain matrix L are selected such that (1) the system (15.7) in the absence of disturbances under control law $u = K_x x$ is asymptotically stable, and (2) the disturbance estimate by DOB can track the disturbance asymptotically.*

Proof Define the state error as

$$e_x = x - x_s.$$

(15.20)

Combining (15.19) and (15.20), the closed-loop system is given by

$$\begin{bmatrix} \dot{e}_x \\ \dot{e}_{d_{lx}} \end{bmatrix} = \begin{bmatrix} A_n + B_n K_x & B_n K_{dx} \\ O & -L \end{bmatrix} \begin{bmatrix} e_x \\ e_{d_{lx}} \end{bmatrix} + \begin{bmatrix} I_{4\times4} + B_n K_{dx} & O \\ O & -I_{4\times4} \end{bmatrix} \begin{bmatrix} d_{lx} - d_{lxs} \\ \dot{d}_{lx} \end{bmatrix}.$$

(15.21)

Similar with the proof of Lemma 15.2, it can be proved that the system matrix of (15.21) is Hurwitz. Assumption 15.1 implies that $\lim_{t\to\infty} \dot{d}_{lx}(t) = 0$ and $\lim_{t\to\infty} [d_{lx}(t) - d_{lxs}] = 0$. To this end, the closed-loop system governed by (15.21) is asymptotically stable. This implies that $\lim_{t\to\infty} e_x(t) = 0$. This completes the proof.

15.4.3 Design of Disturbance Compensation Gain

The main contribution of this work is investigating how to design the disturbance compensation gain K_d such that the effects caused by the "mismatched" disturbances can be attenuated from the output channels asymptotically.

Theorem 15.1 *Suppose that disturbances in system (15.7) satisfy Assumption 15.1 and 15.2, and the feedback control gain K_x and the observer gain matrix L are selected such that $A_n + B_n K_x$ and $-L$ are Hurwitz matrices. The disturbances applied on the system (15.7) can be asymptotically attenuated from the output channel by the newly designed DOBRC law (15.17) if the disturbance compensation gains in K_d are selected as*

$$K_{dx} = [D_n - (C_n + D_n K_x)(A_n + B_n K_x)^{-1} B_n]^{-1}(C_n + D_n K_x)$$
$$\times (A_n + B_n K_x)^{-1} B_{ld}, \tag{15.22}$$

$$K_{dy} = -[D_n - (C_n + D_n K_x)(A_n + B_n K_x)^{-1} B_n]^{-1} D_{ld}. \tag{15.23}$$

Proof By substituting the control law (15.17) into system (15.7), the state is expressed as

$$x = (A_n + B_n K_x)^{-1} \left[\dot{x} - B_n K_d \hat{d} - B_{ld} d_{lx} \right]. \tag{15.24}$$

Combining (15.7), (15.22), and (15.23), with (15.24) gives

$$\begin{aligned}
y &= (C_n + D_n K_x)(A_n + B_n K_x)^{-1} \dot{x} \\
&\quad + (D_n - (C_n + D_n K_x)(A_n + B_n K_x)^{-1} B_n) K_d \hat{d} \\
&\quad - (C_n + D_n K_x)(A_n + B_n K_x)^{-1} B_{ld} d_{lx} + D_{ld} d_{ly} \\
&= (C_n + D_n K_x)(A_n + B_n K_x)^{-1} \\
&\quad \dot{x} + (C_n + D_n K_x)(A_n + B_n K_x)^{-1} B_{ld} e_{d_{lx}} - D_{ld} e_{d_{ly}}.
\end{aligned} \tag{15.25}$$

According to Lemma 15.3, it can be concluded that

$$\lim_{t \to \infty} \dot{x}(t) = 0, \ \lim_{t \to \infty} e_{d_{lx}}(t) = 0. \tag{15.26}$$

It can be deduced form (15.11), (15.25), and (15.26) that

$$\lim_{t \to \infty} y(t) = 0. \tag{15.27}$$

Remark 15.3 *Note that the disturbance compensation gain K_d in (15.22) and (15.23) is a general case and suitable for both "matched" and "mismatched" disturbances. In "matched" case, i.e., $B_n = B_{ld}$ and $D_n = D_{ld} = 0$, it can be obtained from (15.22) and (15.23) that the disturbance compensation gain is reduced to $K_{dx} = -I$ and $K_{dy} = 0$, which is the particular form in the previous literature [86, 2].*

In our work, the observer gain matrix in DO (15.12) is selected as

$$\mathbf{L} = \begin{bmatrix} 50 & 0 & 0 & 0 \\ 0 & 50 & 0 & 0 \\ 0 & 0 & 50 & 0 \\ 0 & 0 & 0 & 50 \end{bmatrix}. \tag{15.28}$$

The disturbance compensation gains can also be calculated by Eqs. (15.22) and (15.23), given as

$$\mathbf{K}_{dx} = \begin{bmatrix} 0.0055 & 0.1883 & 0.0033 & 0.1252 \\ -0.0004 & -0.0407 & 0.0065 & 0.0364 \end{bmatrix}, \tag{15.29}$$

$$\mathbf{K}_{dy} = \begin{bmatrix} -1.232 & 0.1475 \\ -0.0964 & -0.1695 \end{bmatrix}. \tag{15.30}$$

The control structure of the proposed DOBRC for the BTT missile system is shown in Figure 15.1.

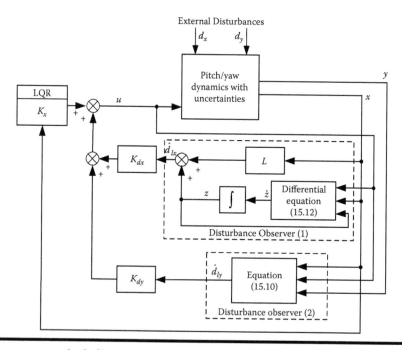

Figure 15.1 **Block diagram of the proposed DOBRC method for the BTT missile system.**

15.5 Simulation Studies

In this section, both external disturbances and model uncertainties are considered to show the effectiveness of the proposed DOBRC methods. In the simulation, it is supposed that the missile travels with a velocity of $v_t = 1000$ m/s and a roll rate of $\omega_x = 400$ deg/s. To demonstrate the efficiency of the proposed method in improving external disturbance rejection performance, and robustness against model uncertainties, the baseline LQR is employed for the purpose of comparison.

15.5.1 External Disturbance Rejection Performance

Suppose that step external disturbances are imposed on each equation in (15.1), actually, $d_{\omega_z} = d_\alpha = d_{\omega_y} = d_\beta = 4$ and $d_{n_z} = d_{n_y} = 1$ at $t = 2$ sec. The output responses and input profiles of the pitch/yaw channels under the proposed DOBRC method are shown in Figures 15.2 and 15.3, respectively. The corresponding responses of the states are shown in Figure 15.4.

As shown in Figures 15.2–15.4, in the first two seconds, the control performance under the proposed method recovered to that under the baseline LQR since there are no disturbances during this time period. It can be observed from Figures 15.2–15.4 that, when the disturbances appear, the feedforward part of the proposed method, which serves as a "patch" to the baseline controller, becomes active in rejecting the disturbances, while the baseline LQR controller fails to effectively counteract the disturbances. This implies that the nominal performance of the proposed method is preserved in the absence of the external disturbances. In the presence of the "mismatched" external disturbances, an excellent disturbance rejection as indicated by a short settling time and offset-free in steady state exhibits by the proposed DOBRC approach. Therefore, a good disturbance rejection ability is achieved without sacrificing the nominal performance, which is one of the major advantages of the proposed method. Furthermore, Figure 15.3 shows that the actuator deflection angles are within the allowable regions, and no excessive control energy is required.

15.5.2 Robustness Against Model Uncertainties

As listed in Table 15.1, the aerodynamic parameters are time-varying during the missile flies. To representing the continuous changes of the real missile dynamics during its flight, a linear transition of the parameters from one operating points to another happens [205] is adopted. For instance, aerodynamic parameters $a_i(t)$, $i =$

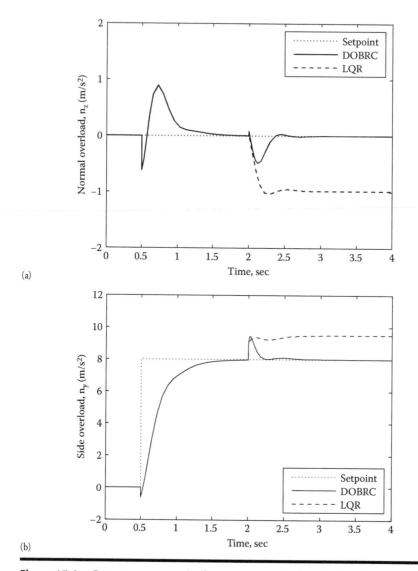

(a)

(b)

Figure 15.2 Output responses in the presence of external disturbances under the proposed DOBRC (solid line) and the LQR (dashed line): (a) normal overload, n_z; (b) side overload, n_y. The reference signals are denoted by dotted lines.

$1, 2, \ldots, 5$, are expressed as

$$
a_i(t) = \begin{cases} a_i(t_1), \ (0 \le t < t_1), \\ a_i(t_{j-1}) + \frac{a_i(t_j) - a_i(t_{j-1})}{t_j - t_{j-1}}(t - t_{j-1}), \\ (t_{j-1} \le t < t_j \ (j = 2, 3, \ldots, 7)). \end{cases} \tag{15.31}
$$

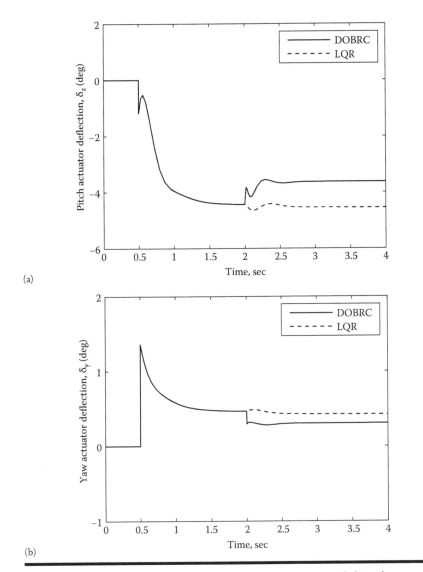

(a)

(b)

Figure 15.3 **Input time histories in the presence of external disturbances under the proposed DOBRC (solid line) and the LQR (dashed line): (a) pitch actuator deflection, δ_z; (b) yaw actuator deflection, δ_y.**

The parameters $b_i(t)$, $i = 1, 2, \ldots, 5$, and $e_i(t)$, $i = 1, 2$, change in a similar form with the time as described by the formula (15.31).

In the presence of such model uncertainties caused by time-varying parameters, the response curves of the outputs, inputs and states under the baseline LQR design and the DOBRC approach are compared in Figures 15.5, 15.5, and 15.6, respectively.

At time $t = 1$ second, a step command is applied on the side overload. Again, both the baseline LQR and the proposed DOBRC cope with this quite well. Due to the coupling between the pitch and yaw channels, a quite significant normal acceleration is generated, but the LQR and DOBRC regulate the normal acceleration to zero as required quite quickly. As the missile undergoes continuous change of its dynamics,

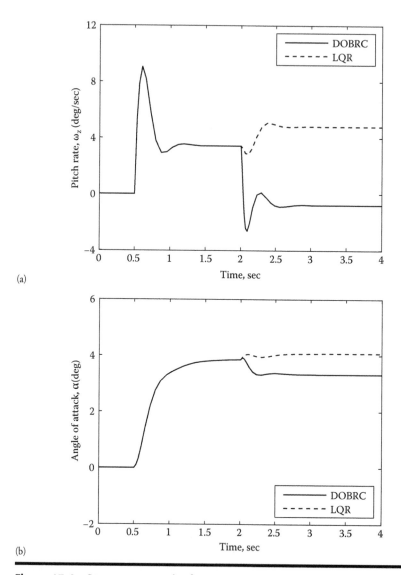

(a)

(b)

Figure 15.4 **State responses in the presence of external disturbances under the proposed DOBRC (solid line) and the LQR (dashed line): (a) pitch rate, ω_z; (b) angle of attack, α; (c) yaw rate, ω_y; (d) sideslip angle, β. (Continued)**

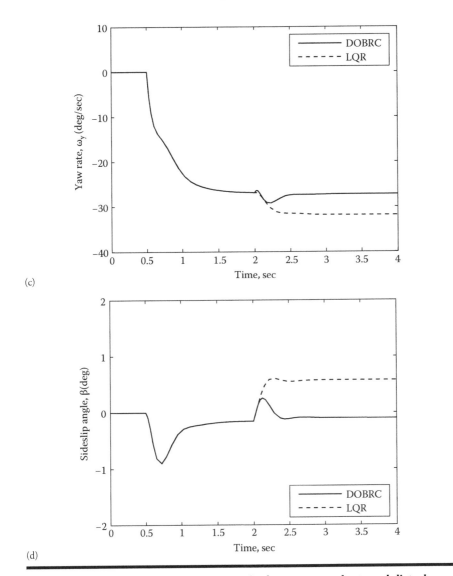

(c)

(d)

Figure 15.4 (*Continued*) **State responses in the presence of external disturbances under the proposed DOBRC (solid line) and the LQR (dashed line): (a) pitch rate, ω_z; (b) angle of attack, α; (c) yaw rate, ω_y; (d) sideslip angle, β.**

gradually a large error has been built up in LQR design, which indicates that the LQR has a good robustness when the model error is within a certain range but poor performance, even instability, may be experienced in the presence of a quite substantial uncertainty. In contrast, the DOBRC approach exhibits an excellent robustness performance.

It can be observed from Figure 15.5 that the output responses can precisely track the reference commands during the whole missile flight regardless of the substantial changes of its parameters. To achieve this, as shown in Figure 15.5, the pitch control δ_z continuously varies to compensate the influence of the parameter variations. Figure 15.6 shows that the yaw rate ω_y experiences a large change during the flight, which

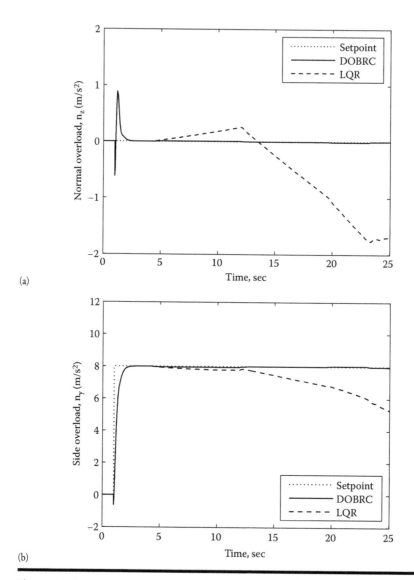

(a)

(b)

Figure 15.5 **Output responses in the presence of model uncertainties under the proposed DOBRC (solid line) and the LQR (dashed line): (a) normal overload, n_z; (b) side overload, n_y. The reference signals are denoted by dotted lines.**

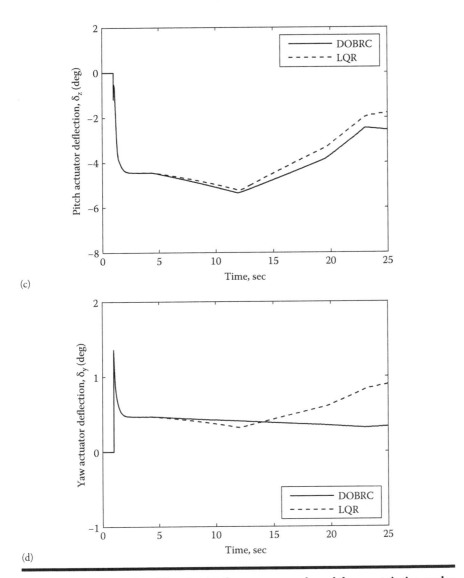

(c)

(d)

Figure 15.5 Input time histories in the presence of model uncertainties under the proposed DOBRC (solid line) and the LQR (dashed line): (a) pitch actuator deflection, δ_z; (b) yaw actuator deflection, δ_y.

may be caused by severe aerodynamic parameter perturbations. It shall be highlighted that the sideslip angle and the actuator deflections are well within the specified ranges. Therefore, it is concluded from this simulation exercise that the baseline LQR exhibits unsatisfactory control performance in this case, while the proposed method achieves surprisingly promising robustness against model uncertainties caused by substantial parameter variations.

The proposed DORBC method provides a promising way to improve the robustness of the baseline controller without resorting to a high gain design as in many robust control methods.

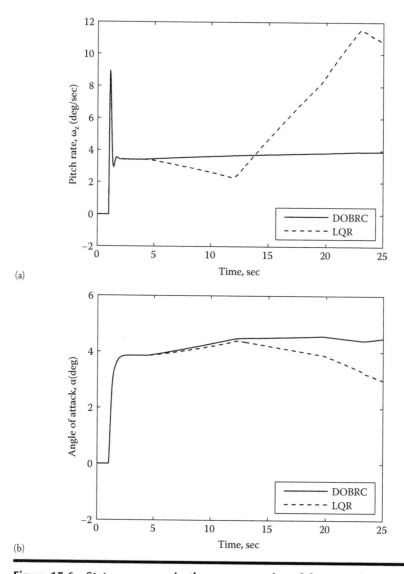

(a)

(b)

Figure 15.6 State responses in the presence of model uncertainties under the proposed DOBRC (solid line) and the LQR (dashed line): (a) pitch rate, ω_z; (b) angle of attack, α; (c) yaw rate, ω_y; (d) sideslip angle, β. (Continued)

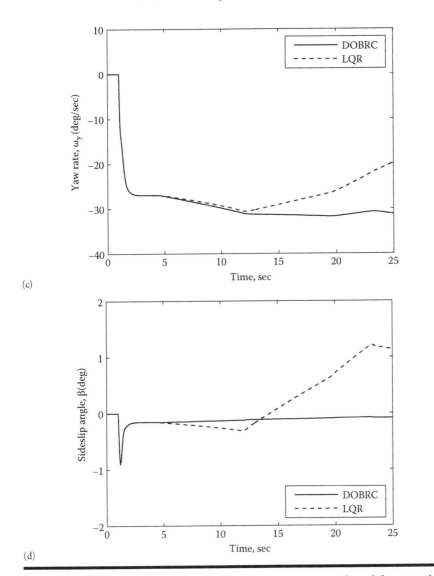

Figure 15.6 *(Continued)* **State responses in the presence of model uncertainties under the proposed DOBRC (solid line) and the LQR (dashed line): (a) pitch rate, ω_z; (b) angle of attack, α; (c) yaw rate, ω_y; (d) sideslip angle, β.**

15.6 Summary

Using a newly proposed disturbance observer based robust control method, the disturbance attenuation problem of a BTT missile has been investigated in this chapter. The "disturbance" under consideration is a general concept, may include both external disturbances and internal disturbances caused by parameter variations. In this setting, the disturbance generally does not satisfy the "matching" condition, i.e., the disturbances enter the system with different channels from the control inputs. A new DOBRC method has been proposed for the BTT missile to improve its disturbance attenuation and in particular the robustness against the substantial variations of the parameters during the flight. It should be pointed out that the proposed method is very easy for practical implementation since only linear analysis and synthesis approaches are employed. The simulation results have demonstrated that very promising disturbance attenuation and strong robustness have been achieved by the proposed disturbance observer-based approach. The results in this chapter show that the disturbance observer-based control technique provides a very promising, practical solution for challenging control problems with large uncertainties and significant external disturbance. Its concept is quite intuitive and practical without resorting to high-gain, nonlinear control theory or other complicated mathematical tools.

Chapter 16

Disturbance Rejection for Airbreathing Hypersonic Vehicles

16.1 Introduction

With the growing interest in developing new technologies in cost efficient space access and speedy global reach, airbreathing hypersonic vehicle (AHV) technique has become more and more significant [56, 57, 206]. The so-called scramjet propulsion is usually employed by the AHV to execute its advantages, such as high speed, cost efficient travel, and large payload [206]. Note that such propulsion mechanism always brings special characteristics of the vehicle dynamics, such as strong nonlinear cross couplings between propulsive and aerodynamic forces [57, 61, 206]. Moreover, the flight of AHV is also significantly influenced by uncertainties caused by the variability of the vehicle characteristics with flight conditions, fuel consumption, and thermal effects on the structure [56, 57, 58, 59, 60, 61, 62, 206]. Another great challenge for the flight control design of AHV is that the longitudinal dynamics are always severely affected by different external disturbances as a result of the complex flight environment [62].

Many elegant control approaches have been employed for the design of guidance and control systems for the longitudinal dynamics of AHVs in the past few years. Based on the models obtained by approximate linearization around specified trim condition or input/output linearization technique, many advanced control methods, such as adaptive control [57], robust control [58, 60, 61, 206], sliding mode control [62], and back-stepping control [207], have been employed for AHVs. Recently, artificial-intelligence-based method has become an active research topic and is widely investigated in literature, such as neural-network-based control [207, 208] and fuzzy-logic-based control [209, 210, 211, 212].

Although these control methods have obtained wide applications and proved to be efficient for AHVs, they mainly focus on the stability or robust stability of the uncertain AHVs. In general, the robustness and disturbance rejection performance of these controllers are achieved at the price of sacrificing the nominal control performance. Most of the above-mentioned methods do not consider active and direct disturbance rejection in the controller design. A novel NDOBC approach is proposed in this chapter to enhance the disturbance rejection performance of the longitudinal dynamics of a generic AHV developed at NASA Langley Research Center [60, 62]. The AHVs under consideration are essentially multi-input multi-output (MIMO) nonlinear systems suffering from mismatched disturbances and uncertainties. This brings big challenges for the NDOBC design since most of the existing NDOBC methods are only applicable for matched disturbance case.

In the proposed framework of this chapter, a finite-time disturbance observer is firstly employed for fast disturbance estimation. A baseline nonlinear dynamic inversion control law is then designed to stabilize the nominal nonlinear dynamics. Finally, a composite NDOBC method is constructed to reject the mismatched disturbances by integrating the disturbance observer and the baseline controller through a disturbance compensation gain. It is shown that the proposed composite controller can eliminate the mismatched disturbances and uncertainties from the output channels of the AHVs by appropriately choosing the disturbance compensation gain matrix.

16.2 Problem Formulation

16.2.1 Longitudinal Dynamics of a Generic AHV

The longitudinal dynamic models for a generic AHV developed at NASA Langley Research Center is taken from [62], depicted by

$$\dot{V} = \frac{T \cos \alpha - D}{m} - \frac{\mu \sin \gamma}{r^2} + d_1, \tag{16.1}$$

$$\dot{\gamma} = \frac{L + T \sin \alpha}{mV} - \frac{(\mu - V^2 r) \cos \gamma}{Vr^2} + d_2, \tag{16.2}$$

$$\dot{h} = V \sin \gamma + d_3, \tag{16.3}$$

$$\dot{\alpha} = q - \dot{\gamma} + d_4, \tag{16.4}$$

$$\dot{q} = \frac{M_{yy}}{I_{yy}} + d_5, \tag{16.5}$$

where

$$L = \frac{1}{2}\rho V^2 S C_L, \ D = \frac{1}{2}\rho V^2 S C_D, \ T = \frac{1}{2}\rho V^2 S C_T,$$

$$M_{yy} = \frac{1}{2}\rho V^2 S\bar{c}\left[C_M(\alpha) + C_M(\delta_e) + C_M(q)\right], \ r = h + R_E.$$

The engine dynamics are modeled by a second-order system

$$\ddot{\beta} = -2\varsigma\omega_n\dot{\beta} - \omega_n^2\beta + \omega_n^2\beta_c + d_6. \tag{16.6}$$

In (16.1)–(16.6), variables V, γ, h, α, and q are velocity, flight path angle, altitude, angle of attack, and pitch rate of the AHV, respectively. L, D, T, M_{yy}, β, and r represent lift, drag, thrust, pitching moment, throttle setting, and radial distance from Earth's center, respectively. Variables $d_i(i = 1, \ldots, 6)$ denote the unknown external disturbances.

The physical and aerodynamic coefficients under this study are simplified around the nominal cruising flight, which is $M = 15$ (the Mach number), $V = 15,060$ ft/s, $h = 110,000$ ft, $\gamma = 0$ rad, and $q = 0$ rad/s [62]. The parametric uncertainty is modeled as an additive perturbation Δ to its nominal values,

$$C_L = 0.6203\alpha, \ C_D = 0.6450\alpha^2 + 0.0043378\alpha + 0.003772,$$

$$C_T = \begin{cases} 0.02576\beta, & \text{if } \beta < 1, \\ 0.0224 + 0.00336\beta, & \text{if } \beta > 1, \end{cases}$$

$$C_M(\alpha) = -0.035\alpha^2 + 0.036617(1 + \Delta C_{M\alpha})\alpha + 5.3261 \times 10^{-6},$$

$$C_M(q) = \frac{\bar{c}}{2V}q(-6.796\alpha^2 + 0.3015\alpha - 0.2289), \ C_M(\delta_e) = c_e(\delta_e - \alpha),$$

$$m = m_0(1 + \Delta m), \ I_{yy} = I_0(1 + \Delta I), \ S = S_0(1 + \Delta S),$$

$$\bar{c} = \bar{c}_0(1 + \Delta\bar{c}), \ \rho = \rho_0(1 + \Delta\rho), \ c_e = 0.0292(1 + \Delta c_e),$$

where C_D, C_L, and C_T represent the drag, lift, and thrust coefficients. $C_M(q)$, $C_M(\alpha)$, and $C_M(\delta_e)$ denote the moment coefficients due to pitch rate, angle of attack, and elevator deflection, respectively. The parameters m, μ, I_{yy}, ρ, S, $\bar{c}$, and R_E are mass, gravitational constant, moment of inertial, density of air, reference area, mean aerodynamic chord, and radius of the Earth, respectively, whose nominal values are listed as follows.

$$m_0 = 9,375 \, slugs, \ \mu = 1.39 \times 10^{16} \, ft^3/s^2, \ I_0 = 7 \times 10^6 \, slugs \cdot ft^2,$$

$$\rho_0 = 0.24325 \times 10^{-4} \, slugs/ft^3, \ S_0 = 3,603 \, ft^2,$$

$$\bar{c}_0 = 80 \, ft, \ R_E = 20,903,500 \, ft.$$

The maximum values of the additive parameter perturbations under consideration are taken as

$$|\Delta m| \leq 0.25, \quad |\Delta I| \leq 0.25, \quad |\Delta S| \leq 0.25, \quad |\Delta \bar{c}| \leq 0.25,$$
$$|\Delta \rho| \leq 0.25, \quad |\Delta c_e| \leq 0.25, \quad |\Delta C_{M_\alpha}| \leq 0.25. \tag{16.7}$$

To show the superiority of the proposed method, severe parameter uncertainties are considered in the later simulation studies. The maximum values of the additive perturbation terms Δ are up to 0.25 as shown in (16.7), which are much larger than those considered in Xu et al. [62]. The control inputs are the throttle setting command β_c and the elevator deflection δ_e, while the outputs are the velocity V and the altitude h, respectively.

16.2.2 Problem Formulation

The AHV dynamic models (16.1)-(16.6) can be rewritten as follows

$$\begin{cases} \dot{x} &= [f(x) + \delta f] + [g(x) + \delta g]u + pd, \\ y &= h(x), \end{cases} \tag{16.8}$$

where the states $x = [x_1, \ldots, x_7]^T = [V, \gamma, h, \alpha, q, \beta, \dot{\beta}]^T$, the controls $u = [u_1, u_2]^T = [\delta_e, \beta_c]^T$, and the outputs $y = [y_1, y_2]^T = [V, h]^T$, respectively. $f(x) = [f_1(x), \ldots, f_7(x)]^T$, $g(x) = [g_1(x), \ldots g_7(x)]^T$, $h(x) = [h_1(x), h_2(x)]^T$ are vector or matrix fields in terms of state x, and $p = [p_1, \ldots, p_6]$. $\delta f = [\delta f_1, \ldots, \delta f_7]^T$ and $\delta g = [\delta g_1, \ldots, \delta g_7]^T$ denote uncertainties caused by physical and aerodynamic parameter perturbations, respectively. Since $\delta f_6 = 0$ and $\delta g_6 = 0$ here for the AHV system, its dynamics can be expressed as follows by lumping the external disturbances and uncertainties together,

$$\begin{cases} \dot{x} &= f(x) + g(x)u + pw, \\ y &= h(x), \end{cases} \tag{16.9}$$

where $w = [w_1, \ldots, w_6]^T = p^T[\delta f + \delta g u] + d$ represents the lumped disturbances. The input relative degrees and disturbance relative degrees [106] of the AHV system are calculated as $(\sigma_1, \sigma_2) = (3, 4)$, and $(\nu_1, \nu_2) = (1, 1)$, respectively. In the absence of disturbances and uncertainties, a new group of coordinate transformation which linearizes the AHV system (16.9) is defined by $z = \Phi(x) = \begin{bmatrix} z_1^T, z_2^T \end{bmatrix}^T$, where

$$z_i = \begin{bmatrix} z_1^i \\ z_2^i \\ \vdots \\ z_{\sigma_i}^i \end{bmatrix} = \begin{bmatrix} h_i(x) \\ L_f h_i(x) \\ \vdots \\ L_f^{\sigma_i - 1} h_i(x) \end{bmatrix},$$

for $i = 1, 2$. The notation L denotes the standard Lie derivative [106]. System (16.9) under the new coordinates is expressed as two subsystems including a velocity and an altitude subsystem, depicted by

$$
S1 : \begin{cases}
\dot{z}_1^1 = z_2^1 + L_p h_1 w, \\
\dot{z}_2^1 = z_3^1 + L_p L_f h_1 w, \\
\dot{z}_3^1 = L_f^3 h_1 + L_g L_f^2 h_1 u + L_p L_f^2 h_1 w, \\
y_1 = h_1 = z_1^1,
\end{cases}
$$

$$
\quad (16.10)
$$

$$
S2 : \begin{cases}
\dot{z}_1^2 = z_2^2 + L_p h_2 w, \\
\dot{z}_2^2 = z_3^2 + L_p L_f h_2 w, \\
\dot{z}_3^2 = z_4^2 + L_p L_f^2 h_2 w, \\
\dot{z}_4^2 = L_f^4 h_2 + L_g L_f^3 h_2 u + L_p L_f^3 h_2 w, \\
y_2 = h_2 = z_1^2.
\end{cases}
$$

It can be observed from (16.10) that the disturbances for the AHV under consideration are mismatched ones.

16.3 Nonlinear Disturbance Observer-Based Robust Flight Control

16.3.1 Composite Control Law Design

To present the main result, a finite time disturbance observer is first addressed as follows.

Assumption 16.1 *The disturbances w in the AHV system (16.9) is twice differentiable, and $\ddot{w}(t)$ has a group of Lipshitz constants $L_i > 0 (i = 1, \ldots, 6)$.*

Since the sixth state equation of the AHV system (16.9) does not affected by the disturbances, we only consider the state equations that contain disturbances when designing the disturbance observer. To do this, let $\bar{x} = p^T x = [x_1, \ldots, x_5, x_7]^T$, gives

$$
\dot{\bar{x}} = \bar{f}(x) + \bar{g}(x) u + w, \quad (16.11)
$$

where $\bar{f}(x) = [f_1(x), \ldots, f_5(x), f_7(x)]^T$, $\bar{g}(x) = [g_1(x), \ldots, g_5(x), g_7(x)]^T$.

A nonlinear disturbance observer (NDO) [8, 132] with finite-time convergence property for estimating the disturbances for the AHV system (16.9) is constructed as

$$
\begin{cases}
\dot{z}_0 = v_0 + \bar{f}(x) + \bar{g}(x)u, \ \dot{z}_1 = v_1, \ \dot{z}_2 = v_2, \\
\hat{x} = z_0, \ \hat{w} = z_1, \ \hat{\dot{w}} = z_2,
\end{cases}
\tag{16.12}
$$

where $z_0 = [z_{01}, \ldots, z_{06}]^T$, $z_1 = [z_{11}, \ldots, z_{16}]^T$, $z_2 = [z_{21}, \ldots, z_{26}]^T$, $v_0 = [v_{01}, \ldots, v_{06}]^T$, $v_1 = [v_{11}, \ldots, v_{16}]^T$, $v_2 = [v_{21}, \ldots, v_{26}]^T$, $\hat{x}$ the estimate of $\bar{x}$, $\hat{w}$ the estimate of w, $\hat{\dot{w}}$ the estimate of $\dot{w}$, and

$$
v_{0i} = -\lambda_0 L_i^{1/3} |z_{0i} - \bar{x}_i|^{2/3} sign(z_{0i} - \bar{x}_i) + z_{1i},
$$

$$
v_{1i} = -\lambda_1 L_i^{1/2} |z_{1i} - v_{0i}|^{1/2} sign(z_{1i} - v_{0i}) + z_{2i},
$$

$$
v_{2i} = -\lambda_2 L_i sign(z_{2i} - v_{1i}),
$$

for $i = 1, \ldots, 6$, $\lambda_0, \lambda_1, \lambda_2 > 0$ are observer coefficients.

Combining (16.11) and (16.12), the ith sub-observer estimation error is given as

$$
\begin{cases}
\dot{e}_{0i} = -\lambda_0 L_i^{1/3} |e_{0i}|^{2/3} sign(e_{0i}) + e_{1i}, \\
\dot{e}_{1i} = -\lambda_1 L_i^{1/2} |e_{1i} - \dot{e}_{0i}|^{1/2} sign(e_{1i} - \dot{e}_{0i}) + e_{2i}, \\
\dot{e}_{2i} = -\lambda_2 L_i sign(e_{2i} - \dot{e}_{1i}) - \ddot{w}_i,
\end{cases}
\tag{16.13}
$$

where the estimation errors are defined as $e_{0i} = z_{0i} - \bar{x}_i$, $e_{1i} = z_{1i} - w_i$, and $e_{2i} = z_{2i} - \dot{w}_i$, respectively. Considering Assumption 16.1, it follows from [8, 132] that system (16.13) is finite-time stable. This implies that there is a time constant t_f such that $e_{0i}(t) = e_{1i}(t) = e_{2i}(t) = 0$ (or equivalently $\hat{x}_i(t) = \bar{x}_i(t)$, $\hat{w}_i(t) = w_i(t)$, $\hat{\dot{w}}_i(t) = \dot{w}_i(t)$) for $t \geq t_f$.

Let $e = [e_0^T, e_1^T, e_2^T]^T$, the dynamics of full observer error are governed by

$$
\dot{e} = \chi(e, \ddot{w}, \lambda_0, \lambda_1, \lambda_2),
\tag{16.14}
$$

which are finite-time stable. A novel robust flight control for the AHV system is presented by the following theorem.

Theorem 16.1 *Consider the AHV system (16.9) under mismatched disturbances satisfying the condition in Assumption 16.1. Also suppose that the observer coefficients $\lambda_0, \lambda_1, \lambda_2$ of the NDO (16.12) is selected such that the observer error (16.13) is finite-time stable. A novel robust flight control law based on NDO which compensates the* *mismatched disturbances from the outputs of the AHV system (16.9) is given by*

$$u = A^{-1}(x)[-B(x) + V(x, y_r) + \Gamma(x)\hat{w}], \tag{16.15}$$

with

$$A(x) = \begin{bmatrix} L_g L_f^2 h_1 \\ L_g L_f^3 h_2 \end{bmatrix}, \quad B(x) = \begin{bmatrix} L_f^3 h_1 \\ L_f^4 h_2 \end{bmatrix},$$

$$V(x, y_r) = \begin{bmatrix} v_1(x, y_r) \\ v_2(x, y_r) \end{bmatrix} = \begin{bmatrix} -c_0^1(h_1 - y_{r1}) - c_1^1 L_f h_1 - c_2^1 L_f^2 h_1 \\ -c_0^2(h_2 - y_{r2}) - c_1^2 L_f h_2 - c_2^2 L_f^2 h_2 - c_3^2 L_f^3 h_2 \end{bmatrix},$$

$$\Gamma(x) = \begin{bmatrix} \gamma_1(x) \\ \gamma_2(x) \end{bmatrix} = \begin{bmatrix} -c_1^1 L_p h_1 - c_2^1 L_p L_f h_1 - L_p L_f^2 h_1 \\ -c_1^2 L_p h_2 - c_2^2 L_p L_f h_2 - c_3^2 L_p L_f^2 h_2 - L_p L_f^3 h_2 \end{bmatrix},$$

where $y_r = [y_{r1}, y_{r2}]^T$, y_{r1} and y_{r2} represent the reference signals of the velocity and altitude, respectively. Parameters $c_k^i (i = 1, 2; k = 0, 1, \ldots, \sigma_i - 1)$ have to be designed such that the polynomials

$$p_0^i(s) = c_0^i + c_1^i s + \cdots + c_{\sigma_i-1}^i s^{\sigma_i-1} + s^{\sigma_i}, \tag{16.16}$$

are Hurwitz stable.

Proof Collecting the last states of the velocity subsystem $\mathcal{S}1$ and altitude subsystem $\mathcal{S}2$ in (16.10) together gives

$$\begin{bmatrix} \dot{z}_3^1 \\ \dot{z}_4^2 \end{bmatrix} = B(x) + A(x)u + \begin{bmatrix} L_p L_f^2 h_1 w \\ L_p L_f^3 h_2 w \end{bmatrix}. \tag{16.17}$$

Substituting control law (16.15) into AHV dynamic system (16.17) yields

$$\begin{bmatrix} \dot{z}_3^1 \\ \dot{z}_4^2 \end{bmatrix} = \begin{bmatrix} v_1(x, y_r) + \gamma_1(x)\hat{w} + L_p L_f^2 h_1 w \\ v_2(x, y_r) + \gamma_2(x)\hat{w} + L_p L_f^3 h_2 w \end{bmatrix}. \tag{16.18}$$

Combining (16.18) with (16.10), the closed-loop system is obtained and given by

$$\begin{cases} \dot{z} = \underline{A}z + \Gamma(x)e_1 + D(x)w + R(y_r), \\ y = Cz \end{cases} \tag{16.19}$$

where $e_1 = \hat{w} - w$, and

$$
A = \left[\begin{array}{ccc|cccc}
0 & 1 & 0 & 0 & 0 & 0 & 0 \\
0 & 0 & 1 & 0 & 0 & 0 & 0 \\
-c_0^1 & -c_1^1 & -c_2^1 & 0 & 0 & 0 & 0 \\
\hline
0 & 0 & 0 & 0 & 1 & 0 & 0 \\
0 & 0 & 0 & 0 & 0 & 1 & 0 \\
0 & 0 & 0 & 0 & 0 & 0 & 1 \\
0 & 0 & 0 & -c_0^2 & -c_1^2 & -c_2^2 & -c_3^2
\end{array}\right], \quad
\Gamma(x) = \left[\begin{array}{c}
0 \\
0 \\
\gamma_1(x) \\
0 \\
0 \\
0 \\
\gamma_2(x)
\end{array}\right],
$$

$$
D(x) = \left[\begin{array}{c}
L_p h_1 \\
L_p L_f h_1 \\
\gamma_1(x) + L_p L_f^2 h_1 \\
L_p h_2 \\
L_p L_f h_2 \\
L_p L_f^2 h_2 \\
\gamma_2(x) + L_p L_f^3 h_2
\end{array}\right], \quad
R(y_r) = \left[\begin{array}{c}
0 \\
0 \\
c_0^1 y_{r1} \\
0 \\
0 \\
c_0^2 y_{r2}
\end{array}\right],
$$

$$
C = \left[\begin{array}{ccc|cccc}
1 & 0 & 0 & 0 & 0 & 0 & 0 \\
0 & 0 & 0 & 1 & 0 & 0 & 0
\end{array}\right].
$$

By simple calculation, it can be verified that

$$
C\underline{A}^{-1} D(x) = 0, \quad C\underline{A}^{-1} R(y_r) = -y_r. \tag{16.20}
$$

It then follows from (16.19) and (16.20) that

$$
y = C\underline{A}^{-1} [\dot{z} - \Gamma(x)e_1 - D(x)w - R(y_r)] = C\underline{A}^{-1} [\dot{z} - \Gamma(x)e_1] + y_r. \tag{16.21}
$$

It can be derived from (16.21) that the disturbances can be decoupled from the output channels in an asymptotical way. Assume that the closed-loop system reaches a bounded steady state and let x_s, e_{1s}, and y_s denote the steady-state values of x, e_1, and y, respectively. The steady-state value y_s can be calculated as

$$
y_s = -C\underline{A}^{-1}\Gamma(x_s)e_{1s} + y_r. \tag{16.22}
$$

Since $e_{1s} = 0$, it follows from (16.22) that $y_s = y_r$.

Remark 16.1 *In the absence of disturbances, if the initial state of the NDO (16.12) is selected as $z_0(0) = \bar{x}(0)$, and $z_1(0) = z_2(0) = 0$, the disturbance estimation by (16.12) satisfies $\hat{w}(t) \equiv 0$. To this end, the control performance under the proposed control law (16.15) recovers to the baseline feedback control law, which implies that the property of nominal performance recovery is retained by the proposed NDOBC.*

16.3.2 Stability Analysis

The stability of the resultant closed-loop system is established by Theorem 16.2.

Theorem 16.2 *AHV system (16.9) under the proposed NDO-based robust flight control law (16.15) is locally input-to-state stable (ISS) if the following conditions are satisfied: (i) the mismatched disturbances satisfy Assumption 16.1; (ii) the control parameters c_k^i are selected such that polynomials $p_0^i(s) = 0$ in (16.16) are Hurwitz stable; (iii) the observer coefficients $\lambda_0, \lambda_1, \lambda_2$ are chosen such that system (16.14) is finite-time stable; (iv) the disturbance compensation gain is designed such that $\Gamma(x)$ and $D(x)$ are continuously differentiable around the operating point.*

Proof Combining (16.14), (16.15) with (16.19), the closed-loop system is obtained and described by

$$\begin{cases} \dot{z} = \underline{A}z + \Gamma(x)e_1 + D(x)w + R(y_r), \\ \dot{e} = \chi(e, \ddot{w}, \lambda_0, \lambda_1, \lambda_2). \end{cases} \quad (16.23)$$

First, the asymptotical stability of the following system

$$\begin{cases} \dot{z} = \underline{A}z + \Gamma(x)e_1, \\ \dot{e} = \chi(e, \ddot{w}, \lambda_0, \lambda_1, \lambda_2), \end{cases} \quad (16.24)$$

is proved. The asymptotical stability of system $\dot{z} = \underline{A}z$ has been guaranteed by condition (ii). Taking into account condition (iv) and consider e_1 as an input of system $\dot{z} = \underline{A}z + \Gamma(x)e_1$, it is concluded from Lemma 5.4 in [113] that $\dot{z} = \underline{A}z + \Gamma(x)e_1$ is locally ISS. Combining this result with condition (iii), it follows from Lemma 5.6 in [113] that system (16.24) is asymptotically stable.

Let $X = [z^T, e^T]^T$, the closed-loop system (16.23) is expressed as

$$\dot{X} = \underline{F}(X) + \underline{G}(X)w + \underline{R}(y_r), \quad (16.25)$$

where

$$\underline{F}(X) = \begin{bmatrix} \underline{A}z + \Gamma(x)e_1 \\ \chi(e, \ddot{w}, \lambda_0, \lambda_1, \lambda_2) \end{bmatrix}, \underline{G}(X) = \begin{bmatrix} D(x) \\ 0 \end{bmatrix}, \underline{R}(y_r) = \begin{bmatrix} R(y_r) \\ 0 \end{bmatrix}.$$

Since system $\dot{X} = \underline{F}(X)$ is asymptotically stable, and $\underline{G}(X)$ is continuously differentiable (which can be derived from condition (iv)), it follows from Lemma 5.4 in [113] that the closed-loop system (16.25) is locally ISS.

16.4 Simulation Studies

The control parameters in this simulation studies are designed as

$$c_0^1 = 0.16, \ c_1^1 = 0.92, \ c_2^1 = 1.7,$$

$$c_0^2 = 0.096, \ c_1^2 = 0.712, \ c_2^2 = 1.94, \ c_3^2 = 2.3.$$

The parameters of the NDO (16.12) is designed as

$$\lambda_0 = 3, \ \lambda_1 = 1.5, \ \lambda_2 = 1.1,$$

$$L_1 = 10, \ L_2 = 0.01, \ L_3 = 100, \ L_4 = 0.5, \ L_5 = 1, \ L_6 = 5.$$

To show the effectiveness of the proposed method in improving the robustness of the baseline controller, the baseline nonlinear dynamic inversion controller is employed in the simulation studies for performance comparison. The tracking commands of the velocity and altitude are 15, 180 ft/s (a step change of 120 ft/s from its nominal flight value 15, 060 ft/s) and 112, 000 ft (a step change of 2,000 ft from its nominal flight value 110, 000 ft), respectively.

16.4.1 External Disturbance Rejection

The unknown external disturbances here are taken as $\omega_1 = 2$, $\omega_3 = -10$, $\omega_6 = -0.2$ at $t = 30$ sec, and $\omega_2 = -0.001 \sin(0.2\pi t)$, $\omega_4 = 0.03$, $\omega_5 = 0.1 - 0.1 \sin(0.3\pi t + \pi)$ at $t = 60$ sec for the AHV system. Response curves of the outputs, inputs and states of the AHV system under two controller are shown by Figure 16.1. The external disturbances and their estimations are shown by Figure 16.2.

It is observed from Figure 16.1 that the baseline controller leads to undesirable control performance in the presence of disturbances, while the proposed NDOBC obtains fine external disturbance rejection properties. Figure 16.2 shows that the disturbance estimates under the finite-time disturbance observer converge to their real values in finite time.

Figure 16.1 shows that response curves of the variables under the proposed NDOBC method are the same as those under the baseline control method during the first 50 seconds when there are no disturbances imposed on the system. This implies that the nominal performance under the proposed method is retained.

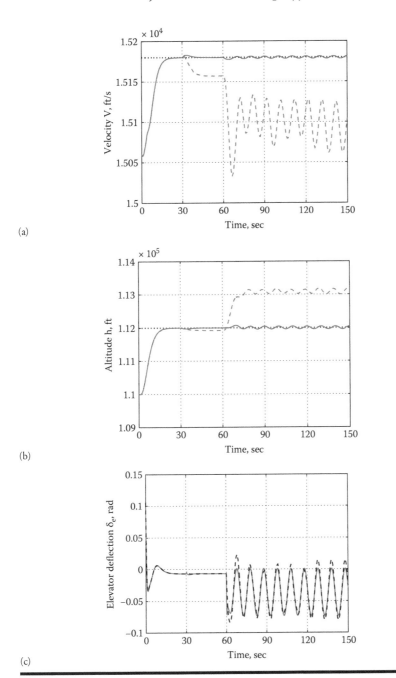

(a)

(b)

(c)

Figure 16.1 **Variable response curves in the presence of unknown external disturbances under the proposed NDOBC (solid line) and baseline control (dashed line) methods. The tracking commands are denoted by dash-dotted line.** *(Continued)*

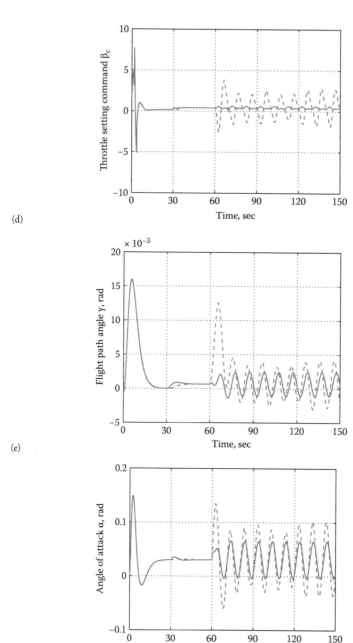

(d)

(e)

(f)

Figure 16.1 *(Continued)* **Variable response curves in the presence of unknown external disturbances under the proposed NDOBC (solid line) and baseline control (dashed line) methods. The tracking commands are denoted by dash-dotted line.** *(Continued)*

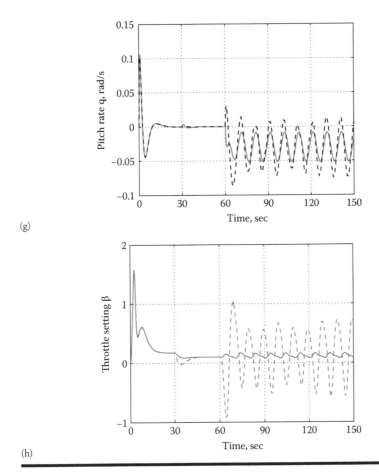

(g)

(h)

Figure 16.1 *(Continued)* **Variable response curves in the presence of unknown external disturbances under the proposed NDOBC (solid line) and baseline control (dashed line) methods. The tracking commands are denoted by dash-dotted line.**

16.4.2 Robustness Against Parameter Uncertainties

Two cases of parameter uncertainties are taken into account, and actually the uncertain parameters (16.7) are taken as their maximum values.

Case I: *All the uncertain parameters (16.7) are taken as their positive maximum values, i.e.,* $\Delta = 0.25$.

The response curves of all variables of the AHV system in the presence of model uncertainties in Case I are shown in Figure 16.3. It is shown by Figure 16.3 that the baseline control method leads to a poor tracking performance, while the outputs under the proposed method can track their commands timely and without any steady state errors.

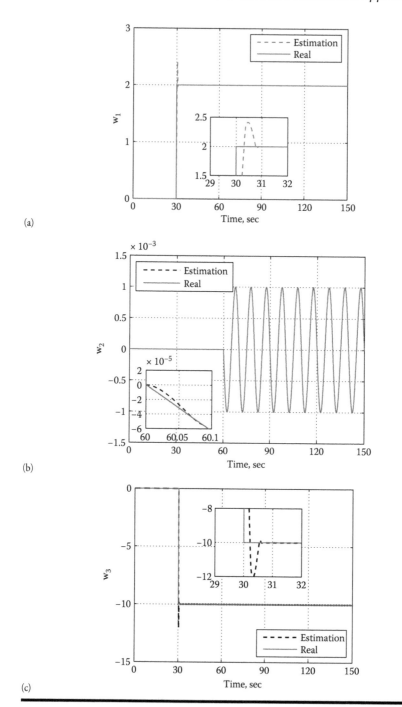

Figure 16.2 Response curves of disturbance estimations in the presence of un-known external disturbances. (Continued)

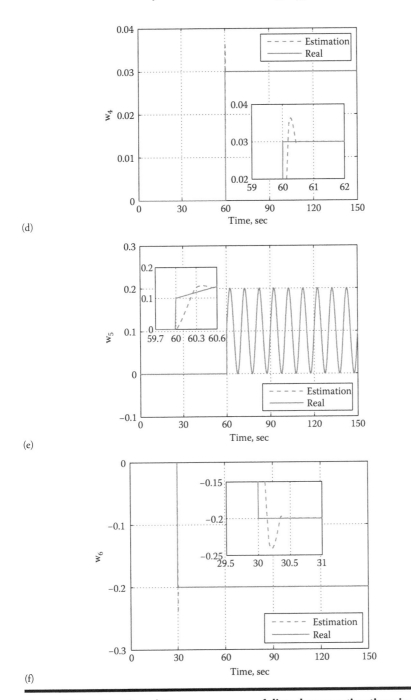

(d)

(e)

(f)

Figure 16.2 *(Continued)* **Response curves of disturbance estimations in the presence of unknown external disturbances.**

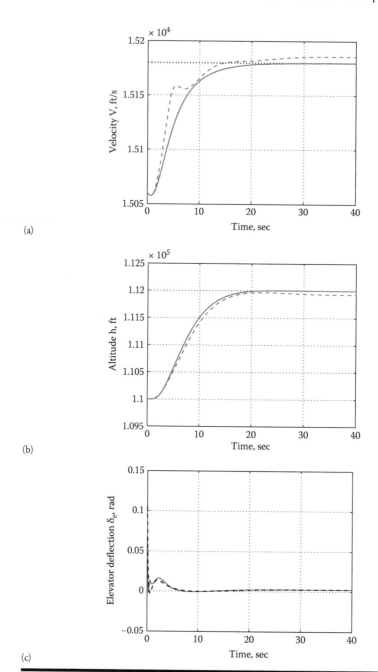

(a)

(b)

(c)

Figure 16.3 Variable response curves in the presence of parameter uncertainties (Case I) under the proposed NDOBC (solid line) and baseline control (dashed line) methods. The tracking commands are denoted by dash-dotted line. *(Continued)*

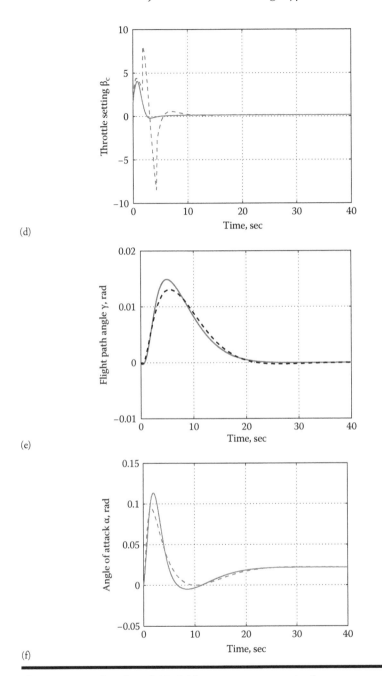

(d)

(e)

(f)

Figure 16.3 *(Continued)* Variable response curves in the presence of parameter uncertainties (Case I) under the proposed NDOBC (solid line) and baseline control (dashed line) methods. The tracking commands are denoted by dash-dotted line. *(Continued)*

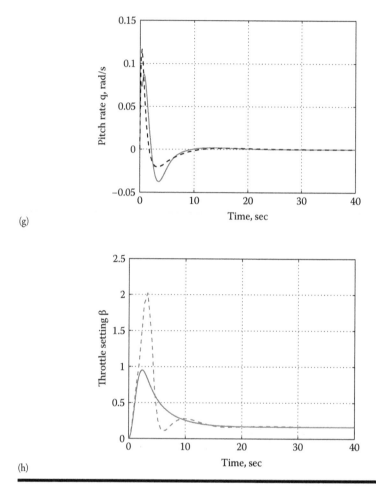

(g)

(h)

Figure 16.3 *(Continued)* **Variable response curves in the presence of parameter uncertainties (Case I) under the proposed NDOBC (solid line) and baseline control (dashed line) methods. The tracking commands are denoted by dash-dotted line.**

Case II: *All the uncertain parameters (16.7) are taken as their negative maximum values, i.e.,* $\Delta = -0.25$.

The response curves of all variables of the AHV system in the presence of model uncertainties in Case II are shown in Figure 16.4. As shown by Figure 16.4, the baseline controller results in a quite poor tracking performance. However, Figure 16.4 shows that the proposed NDOBC method achieves excellent robust tracking performance in such case of severe model uncertainties.

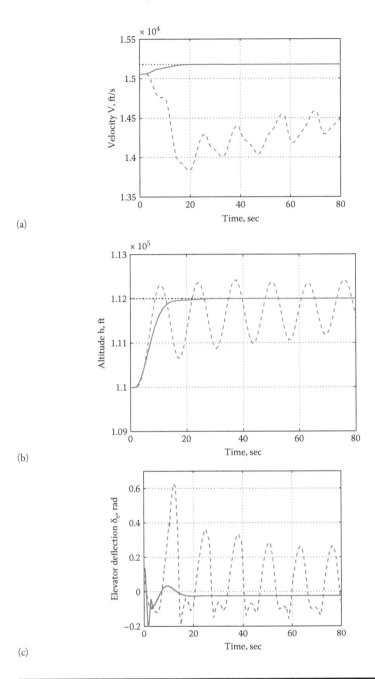

(a)

(b)

(c)

Figure 16.4   Variable response curves in the presence of parameter uncertainties (Case II) under the proposed NDOBC (solid line) and baseline control (dashed line) methods. The tracking commands are denoted by dash-dotted line. *(Continued)*

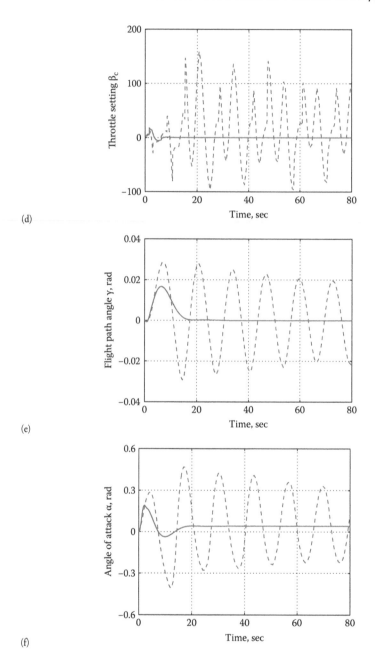

(d)

(e)

(f)

Figure 16.4 *(Continued)* Variable response curves in the presence of parameter uncertainties (Case II) under the proposed NDOBC (solid line) and baseline control (dashed line) methods. The tracking commands are denoted by dash-dotted line. *(Continued)*

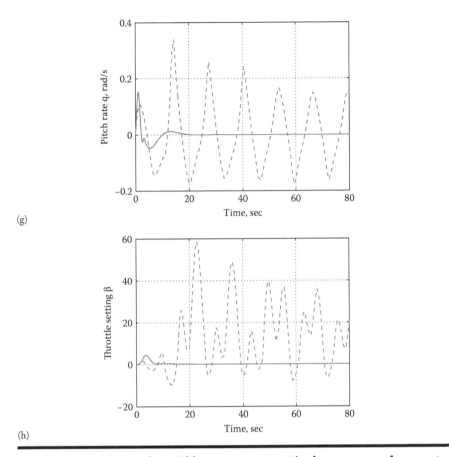

(g)

(h)

Figure 16.4 *(Continued)* **Variable response curves in the presence of parameter uncertainties (Case II) under the proposed NDOBC (solid line) and baseline control (dashed line) methods. The tracking commands are denoted by dash-dotted line.**

16.5 Summary

A novel robust flight control design method has been proposed for the longitudinal dynamic models of a generic AHV system via a NDOBC method. It has been shown that the mismatched uncertainties in the AHV system can be removed from the outputs by the proposed method with properly chosen disturbance compensation gain matrix. The proposed method has obtains not only good robustness against mismatched disturbances and uncertainties but also the property of nominal control performance recovery.

References

[1]   J. Yang, S.H. Li, X.S. Chen, and Q. Li. Disturbance rejection of ball mill grinding circuits using DOB and MPC. *Powder Technology*, 198(2):219–228, 2010.

[2]   W.H. Chen, D.J. Ballance, P.J. Gawthrop, and J. O'Reilly. A nonlinear disturbance observer for robotic manipulators. *IEEE Transactions on Industrial Electronics*, 47(4):932–938, 2000.

[3]   W.H. Chen. Nonlinear disturbance observer-enhanced dynamic inversion control of missiles. *Journal of Guidance, Control and Dynamics*, 26(1):161–166, 2003.

[4]   J. Han. From PID to active disturbance rejection control. *IEEE Transactions on Industrial Electronics*, 56(3):900–906, 2009.

[5]   W.H. Chen. Disturbance observer based control for nonlinear systems. *IEEE/ASME Transactions on Mechatronics*, 9(4):706–710, 2004.

[6]   W.H. Chen. Harmonic disturbance observer for nonlinear systems. *Journal of Dynamic Systems Measurement and Control*, 125(1):114–117, 2003.

[7]   L.B. Freidovich and H.K. Khalil. Performance recovery of feedback-linearization-based designs. *IEEE Transactions on Automatic Control*, 53(10):2324–2334, 2008.

[8]   A. Levant. Higher-order sliding modes, differentiation and output-feedback control. *International Journal of Control*, 76(9/10):924–941, 2003.

[9]   J. Han. Extended state observer for a class of uncertain plants. *Control and Decision*, 10(1):85–88, 1995 (in Chinese).

[10] W.H. Chen, D.J. Ballance, P.J. Gawthrop, J.J. Gribble, and J. O'Reilly. Nonlinear PID predictive controller. *IEE Proceedings Control Theory and Applications*, 146(6):603–611, 1999.

[11] K.S. Kim, K.H. Rew, and S. Kim. Disturbance observer for estimating higher order disturbances in time series expansion. *IEEE Transactions on Automatic Control*, 55(8):1905–1911, 2010.

[12] J. Yang, A. Zolotas, W.H. Chen, K. Michail, and S.H. Li. Robust control of nonlinear MAGLEV suspension system with mismatched uncertainties via DOBC approach. *ISA Transactions*, 3(50):389–396, 2011.

[13] J. Yang, W.H. Chen, and S.H. Li. Autopilot design of bank-to-turn missile using state-space disturbance observers. In *UKACC International Conference on Control 2010*, pages 1218–1223, 2010.

[14] S.H. Li, J. Yang, W.H. Chen, and X.S. Chen. Generalized extended state observer based control for systems with mismatched uncertainties. *IEEE Transactions on Industrial Electronics*, 59(12):4792–4802, 2012.

[15] J. Yang, W.H. Chen, and S.H. Li. Non-linear disturbance observer-based robust control for systems with mismatched disturbances/uncertainties. *IET Control Theory and Application*, 5(18):2053–2062, 2011.

[16] J. Yang, W.H. Chen, S.H. Li, and X.S. Chen. Static disturbance-to-output decoupling for nonlinear systems with arbitrary disturbance relative degree. *International Journal of Robust and Nonlinear Control*, Published online.

[17] J. Yang, S.H. Li, and W.H. Chen. Nonlinear disturbance observer-based control for multi-input multi-output nonlinear systems subject to mismatching condition. *International Journal of Control*, 8(85):1071–1082, 2012.

[18] J. Yang, S.H. Li, and X.H. Yu. Sliding-mode control for systems with mismatched uncertainties via a disturbance observer. *IEEE Transactions on Industrial Electronics*, 60(1):160–169, 2013.

[19] X.K. Chen, T. Fukuda, and K.D. Young. A new nonlinear robust disturbance observer. *Proceedings of the Institution of Mechanical Engineers Part G, Journal of Aerospace Engineering*, 211(G3):169–181, 2000.

[20] X.S. Chen, J. Yang, S.H. Li, and Q. Li. Disturbance observer based multivariable control of ball mill grinding circuits. *Journal of Process Control*, 19(7):1025–1031, 2009.

[21] O. Galan, G.W. Barton, and J.A. Romagnoli. Robust control of a SAG mill. *Powder Technology*, 124(3):264–271, 2002.

[22] K. Najim, D. Hodouin, and A. Desbiens. Adaptive control: State of the art and application to a grinding process. *Powder Technology*, 82(1):59–68, 1995.

[23] H. Oh and W. Chung. Disturbance-observer-based motion control of redundant manipulators using inertially decoupled dynamics. *IEEE Transactions on Mechatronics*, 4(2):133–146, 1999.

[24] S. Komada. Control of redundant manipulators considering order of disturbance observer. *IEEE Transactions on Industrial Electronics*, 47(2):413–420, 2000.

[25] K.S. Eom, H. Suh, and W.K. Chung. Disturbance observer based path tracking control of robot manipulator considering torque saturation. *Mechatronics*, 11(3):325–343, 2001.

[26] T. Umeno, T. Kaneko, and Y. Hori. Robust servosystem design with two degrees of freedom and its application to novel motion control of robot manipulators. *IEEE Transactions on Industrial Electronics*, 40(5):473–485, 1993.

[27] K. Ohishi, M. Nakao, K. Ohnishi, and K. Miyachi. Microprocessor-controlled DC motor for load-insensitive position servo system. *IEEE Transactions on Industrial Electronics*, 34(1):44–49, 1987.

[28] T. Umeno and Y. Hori. Robust speed control of DC servomotors using modern two degrees-of-freedom controller design. *IEEE Transactions on Industrial Electronics*, 38(5):363–368, 1991.

[29] L. Guo and M. Tomizuka. High-speed and high-precision motion control with an optimal hybrid feedforward controller. *IEEE Transactions on Mechatronics*, 2(2):110–122, 1997.

[30] M. Bertoluzzo, G.S. Buja, and E. Stampacchia. Performance analysis of a high-bandwidth torque disturbance compensator. *IEEE Transactions on Mechatronics*, 9(4):653–660, 2004.

[31] Y.S. Lu. Smooth speed control of motor drives with asymptotic disturbance compensation. *Control Engineering Practice*, 16(5):597–608, 2008.

[32] D. Cho, Y. Kato, and D. Spilman. Sliding mode and classical controller in magnetic levitation systems. *IEEE Control System Magazine*, 13(1):42–48, 1993.

[33] P.K. Sinha and A.N. Pechev. Model reference adaptive control of a maglev system with stable maximum descent criterion. *Automatica*, 35(8):1457–1465, 1999.

[34] A. Bittar and R.M. Sales. H-2 and H-infinity control for MagLev vehicles. *IEEE Control System Magazine*, 18(4):18–25, 1998.

[35] P.K. Sinha and A.N. Pechev. Nonlinear H-infinity controllers for electromagnetic suspension systems. *IEEE Transactions on Automatic Control*, 2004(4):563–568, 2004.

[36] I. Takahashi and Y. Ide. Decoupling control of thrust and attractive force of a LIM using a space vector control inverter. *IEEE Transactions on Industrial Applications*, 29(1):161–167, 1993.

[37] C. MacLeod and R.M. Goodall. Frequency-shaping LQ control of Maglev suspension systems for optimal performance with deterministic and stochastic inputs. *IEE Proceedings-Control Theory and Application*, 143(1):25–30, 1996.

[38] Y. Huang and W. Messner. A novel disturbance observer design for magnetic hard drive servo system with a rotary actuator. *IEEE Transactions on Magnetics*, 34(4):1892–1894, 1998.

[39] J. Ishikawa. A novel add-on compensator for cancelation of pivot nonlinearities in hard disk drives. *IEEE Transactions on Magnetics*, 34(4):1895–1897, 1998.

[40] W H. Chen. Nonlinear disturbance observer-enhanced dynamic inversion control of missiles. *Journal of Guidance Control and Dynamics*, 26(1):161–166, 2003.

[41] L.C. Fu, W.D. Chang, J.H. Yang, and T.S. Kuo. Adaptive robust bank-to-turn missile autopilot design using neural networks. *Journal of Guidance Control and Dynamics*, 20(2):346–354, 1997.

[42] C.F. Lin, J.R. Cloutier, and J.H. Evers. High-performance, robust, bank-to-turn missile autopilot design. *Journal of Guidance Control and Dynamics*, 18(1):46–53, 1995.

[43] C.K. Lin. Mixed H_2/H_∞ autopilot design of bank-to-turn missiles using fuzzy basis function networks. *Fuzzy Sets and Systems*, 158(20):2268–2287, 2007.

[44] R.A. Nichols, R.T. Reichert, and W.J. Rugh. Gain scheduling for H-infinity controllers: A flight control example. *IEEE Transactions on Control Systems Technology*, 1(2):69–79, 1993.

[45] L.H. Carter and J.S. Shamma. Gain-scheduled bank-to-turn autopilot design using linear parameter varying transformations. *Journal of Guidance Control and Dynamics*, 19(5):1056–1063, 1996.

[46] C.H. Lee, M.H. Shin, and M.J. Chung. A design of gain-scheduled control for a linear parameter varying system: An application to flight control. *Control Engineering Practice*, 9(1):11–21, 2001.

[47] F. Tan and G.R. Duan. Global stabilizing controller design for linear time-varying systems and its application on BTT missiles. *Journal of Systems Engineering and Electronics*, 19(6):1178–1184, 2008.

[48] C.H. Lee and M.J. Chung. Gain-scheduled state feedback control design technique for flight vehicles. *IEEE Transactions on Aerospace and Electronic Systems*, 37(1):173–182, 2001.

[49] M.J. Kim, W.H. Kwon, and Y.H. Kim. Autopilot design for bank-to-turn missiles using receding horizon predictive control scheme. *Journal of Guidance Control and Dynamics*, 20(6):1248–1254, 1997.

[50] C.K. Lin. Adaptive critic autopilot design of bank-to-turn missiles using fuzzy basis function networks. *IEEE Transactions on Systems, Man, and Cybernetics-Part B: Cybernetics*, 35(2):197–207, 2005.

[51] C.K. Lin and S.D. Wang. A self-organizing fuzzy control approach for bank-to-turn missiles. *Fuzzy Sets and Systems*, 96(3):281–306, 1998.

[52] H.J. Uang and B.S. Chen. Robust adaptive optimal tracking design for uncertain missile systems: A fuzzy approach. *Fuzzy Sets and Systems*, 126(1):63–87, 2002.

[53] C.K. Lin and S.D. Wang. An adaptive H_∞ controller design for bank-to-turn missiles using ridge Gaussian neural networks. *IEEE Transactions on Neural Networks*, 15(6):1507–1516, 2004.

[54] D.M. McDowell, G.W. Irwin, G. Lightbody, and G. McConnell. Hybrid neural adaptive control for bank-to-turn missiles. *IEEE Transactions on Control Systems Technology*, 5(3):297–308, 1997.

[55] M.B. McFarland and A.J. Calise. Adaptive nonlinear control of agile antiair missiles using neural networks. *IEEE Transactions on Control Systems Technology*, 8(5):749–756, 2000.

[56] M.A. Bolender and D.B. Doman. Nonlinear longitudinal dynamical model of an air-breathing hypersonic vehicle. *Journal of Spacecraft and Rockets*, 44(2):374–387, 2007.

[57] L. Fiorentini and A. Serrani. Nonlinear robust adaptive control of flexible air-breathing hypersonic vehicles. *Journal of Guidance Control and Dynamics*, 32(2):401–416, 2009.

[58] C.I. Marrison and R.F. Stengel. Design of robust control systems for a hypersonic aircraft. *Journal of Guidance Control and Dynamics*, 21(1):58–63, 1998.

[59] J.T. Parker, A. Serrani, S. Yurkovich, M.A. Bolender, and D.B. Doman. Control-oriented modeling of an airbreathing hypersonic vehicle. *Journal of Guidance Control and Dynamics*, 30(3):856–869, 2007.

[60] Q. Wang and R.F. Stengel. Robust nonlinear control of a hypersonic aircraft. *Journal of Guidance Control and Dynamics*, 23(4):577–585, 2000.

[61] Z.D. Wilcox, W. MacKunis, S. Bhat, R. Lind, and W.E. Dixon. Lyapunov-based exponential tracking control of a hypersonic aircraft with aerothermoelastic effects. *Journal of Guidance Control and Dynamics*, 33(4):1213–1224, 2010.

[62] H.J. Xu, M.D. Mirmirani, and P.A. Ioannou. Adaptive sliding mode control design for a hypersonic flight vehicle. *Journal of Guidance Control and Dynamics*, 27(5):829–838, 2004.

[63] H. Yoon and B.N. Agrawal. Adaptive control of uncertain Hamiltonian multi-input multi-output systems: With application to spacecraft control. *IEEE Transactions on Control Systems Technology*, 17(4):900–906, 2009.

[64] Y. Park. Robust and optimal attitude stabilization of spacecraft with external disturbances. *Aerospace Science and Technology*, 9(3):253–259, 2008.

[65] I. Ali, G. Radice, and J. Kim. Backstepping control design with actuator torque bound for spacecraft attitude maneuver. *Journal of Guidance Control and Dynamics*, 33(1):254–259, 2010.

[66] Q. Zheng and F. Wu. Nonlinear H_∞ control designs with axisymmetric spacecraft control. *Journal of Guidance Control and Dynamics*, 32(3):957–963, 2009.

[67] S.H. Ding and S.H. Li. Stabilization of the attitude of a rigid spacecraft with external disturbances using finite-time control technique. *Aerospace Science and Technology*, 13(4-5):256–265, 2009.

[68] S.H. Li, S.H. Ding, and Q. Li. Global set stabilization of the spacecraft attitude control problem based on quaternion. *International Journal of Robust and Nonlinear Control*, 20(1):84–105, 2010.

[69] C. Johnson. Accommodation of external disturbances in linear regulator and servomechanism problems. *IEEE Transactions on Automatic Control*, 16(6):635–644, 1971.

[70] S. Kwon and W.K. Chung. Robust performance of a multiloop perturbation compensator. *IEEE Transactions on Mechatronics*, 7(2):190–200, 2002.

[71] S. Kwon and W.K. Chung. A discrete-time design and analysis of perturbation observer for motion control applications. *IEEE Transactions on Control System Technology*, 11(3):399–407, 2003.

[72] J.H. She, M. Fang, Y. Ohyama, H. Hashimoto, and M. Wu. Improving disturbance-rejection performance based on an equivalent-input-disturbance approach. *IEEE Transactions on Industrial Electronics*, 55(1):380–389, 2008.

[73] J.H. She, X. Xin, and Y. Pan. Equivalent-input-disturbance approach—analysis and application to disturbance rejection in dual-stage feed drive control system. *IEEE Transactions on Mechatronics*, 16(2):330–340, 2011.

[74] Z. Gao, Y. Huang, and J. Han. An alternative paradigm for control system design. In *Proceedings of 40th IEEE Conference on Decision and Control*, pages 4578–4585, 2001.

[75] D. Sun. Comments on active disturbance rejection control. *IEEE Transactions on Industrial Electronics*, 54(6):3428–3429, 2007.

[76] Y. Xia, P. Shi, G.P. Liu, D. Rees, and J. Han. Active disturbance rejection control for uncertain multivariable systems with time-delay. *IET Control Theory and Applications*, 1(1):75–81, 2007.

[77] Y. Choi, K. Yang, and W.K. Chung. On the robustness and performance of disturbance observers for second-order systems. *IEEE Transactions on Automatic Control*, 48(2):315–320, 2003.

[78] H. Shim and N.H. Jo. An almost necessary and sufficient condition for robust stability of closed-loop systems with disturbance observer. *Automatica*, 45(1):296–299, 2009.

[79] K. Yamada, S. Komada, M. Ishida, and T. Hori. Analysis and classical control design of servo system using high order disturbance observer. In *Proceedings of IEEE International Conference on Industrial Electronics Control and Instrumentation*, pages 4–9, 1997.

[80] S.M. Shahruz. Performance enhancement of a class of nonlinear systems by disturbance observers. *IEEE Transactions on Mechatronics*, 5(3):319–323, 2000.

[81] C.J. Kempf and S. Kobayashi. Disturbance observer and feedforward design for a high-speed direct-drive position table. *IEEE Transactions on Control Systems Technology*, 7(5):513–526, 1999.

[82] B.A. Guvenc and L. Guvenc. Robust two degree-of-freedom add-on controller design for automatic steering. *IEEE Transactions on Control Systems Technology*, 10(1):137–148, 2002.

[83] M. Defoort and T. Murakami. Second order sliding mode control with disturbance observer for bicycle stabilization. In *Proceedings of IEEE/RSJ International Conference on Intelligent Robots and Systems*, pages 2822–2827, 2008.

[84] Z.J. Yang, H. Tsubakihara, S. Kanae, K. Wada, and C.Y. Su. A novel robust nonlinear motion controller with disturbance observer. *IEEE Transactions on Control Systems Technology*, 16(1):137–147, 2008.

[85] X.K. Chen, S. Komada, and T. Fukuda. Design of a nonlinear disturbance observer. *IEEE Transactions on Industrial Electronics*, 47(2):429–437, 2000.

[86] L. Guo and W.H. Chen. Disturbance attenuation and rejection for systems with nonlinearity via DOBC approach. *International Journal of Robust and Nonlinear Control*, 15(3):109–125, 2005.

[87] E. Kim. A fuzzy disturbance observer and its application to control. *IEEE Transactions on Fuzzy Systems*, 10(1):77–85, 2002.

[88] E. Kim. A discrete-time fuzzy disturbance observer and its application to control. *IEEE Transactions on Fuzzy Systems*, 11(3):399–410, 2002.

[89] J.S. Co and B.M. Han. Precision position control of PMSM using neural network disturbance observer and parameter compensator. In *Proceedings of IEEE International Conference on Mechatronics*, pages 316–320, 2006.

[90] H. Shim and Y.J. Joo. State space analysis of disturbance observer and a robust stability condition. In *Proceedings of the 46th IEEE Conference on Decision and Control*, pages 759–764, 2007.

[91] J. Back and H. Shim. Adding robustness to nominal output-feedback controllers for uncertain nonlinear systems: A nonlinear version of disturbance observer. *Automatica*, 44(10):2528–2537, 2008.

[92] J. Back and H. Shim. An inner-loop controller guaranteeing robust transient performance for the uncertain MIMO nonlinear systems. *IEEE Transactions on Automatic Control*, 54(7):1601–1607, 2009.

[93] X.J. Wei and L. Guo. Composite disturbance-observer-based control and H-infinity control for complex continuous models. *International Journal of Robust and Nonlinear Control*, 20(1):106–118, 2010.

[94] X.J. Wei, H.F. Zhang, and L. Guo. Composite disturbance-observer-based control and terminal sliding mode control for uncertain structural systems. *International Journal of Systems Science*, 40(10):1009–1017, 2009.

[95] X.J. Wei, H.F. Zhang, and L. Guo. Saturating composite disturbance-observer-based control and H-infinity control for discrete time-delay systems with nonlinearity. *International Journal of Control Automation and Systems*, 7(5):691–701, 2009.

[96] X.J. Wei, H.F. Zhang, and L. Guo. Composite disturbance-observer-based control and variable structure control for non-linear systems with disturbances. *Transactions of the Institute of Measurement and Control*, 31(5):401–423, 2009.

[97] X.J. Wei and L. Guo. Composite disturbance-observer-based control and terminal sliding mode control for non-linear systems with disturbances. *International Journal of Control*, 92(6):1082–1098, 2009.

[98] M. Chen and W.H. Chen. Disturbance-observer-based robust control for time delay uncertain systems. *International Journal of Control Automation and Systems*, 8(2):445–453, 2010.

[99] M. Chen and W.H. Chen. Sliding mode control for a class of uncertain nonlinear system based on disturbance observer. *International Journal of Adaptive Control and Signal Processing*, 24(1):51–64, 2010.

[100] B.R. Barmish and G. Leitmann. On ultimate boundness control of uncertain systems in the absence of matching assumption. *IEEE Transactions on Automatic Control*, 27(1):153–158, 1982.

[101] R. Errouissi and M. Ouhrouche. Nonlinear predictive controller for a permanent magnet synchronous motor drive. *Mathematics and Computers in Simulation*, 81(2):394–406, 2010.

[102] Y.A.R.I. Mohamed. Design and implementation of a robust current-control scheme for a PMSM vector drive with a simple adaptive disturbance observer. *IEEE Transactions on Industrial Electronics*, 54(4):1981–1988, 2007.

[103] K. Michail. *Optimised Configuration of Sensing for Control and Fault Tolerance Applied to an Elctromagnetic Suspension System*. Loughborough University, Leicester, England, 2009.

[104] B.W. Bequette. *Process Control: Modeling, Design, and Simulation*. Prentice Hall, New Jersey, 2002.

[105] H.D. Zhu, G.H. Zhang, and H.H. Shao. Control of the process with inverse response and dead-time based on disturbance observer. In *Proceedings of American Control Conference*, pages 4826–4831, 2005.

[106] A. Isidori. *Nonlinear Control Systems: An Introduction, 3rd ed.* Springer-Verlag, 1995.

[107] A. Isidori and C.J. Byrnes. Output regulation of nonlinear systems. *IEEE Transactions on Automatic Control*, 2(35):131–140, 1990.

[108] C.E. Hall and Y.B. Shtessel. Sliding mode disturbance observer-based control for a reusable launch vehicle. *Journal of Guidance Control and Dynamics*, 29(6):1315–1328, 2006.

[109] Y.S. Lu. Sliding-mode disturbance observer with switching-gain adaptation and its application to optical disk drives. *IEEE Transactions on Industrial Electronics*, 56(9):3743–3750, 2009.

[110] Q. Zheng, L.Q. Gao, and Z. Gao. On stability analysis of active disturbance rejection control for nonlinear time-varying plants with unknown dynamics. In *Proceedings of 46th IEEE Conference on Decision and Control*, pages 3501–3506, 2007.

[111] W.H. Chen and L. Guo. Analysis of disturbance observer based control for nonlinear systems under disturbances with bounded variation. In *Proceedings of UKACC 2004*, 2004.

[112] Z. Gao. Active disturbance rejection control: A paradigm shift in feedback control system design. In *Proceedings of American Control Conference*, pages 2399–2405, 2006.

[113] H.K. Khalil. *Nonlinear Systems, 2nd ed.* Prentice-Hall, 1996.

[114] R. Reichert. Modern robust control for missile autopilot design. In *Proceedings of American Control Conference*, pages 2368–2373, 1990.

[115] P. Lu. Nonlinear predictive controllers for continuous systems. *Journal of Guidance Control and Dynamics*, 17(3):553–560, 1994.

[116] W.H. Chen, D.J. Ballance, and P.J. Gawthrop. Optimal control of nonlinear systems: a predictive control approach. *Automatica*, 39(4):633–641, 2003.

[117] S.H. Lane and R.F. Stengel. Flight control design using non-linear inverse dynamics. *Automatica*, 24(4):471–483, 1988.

[118] C. Schumacher and P.P. Khargonekar. Missile autopilot designs using H_∞ control with gain scheduling and dynamic inversion. *Journal of Guidance Control and Dynamics*, 21(2):234–243, 1988.

[119] A. Isidori. *Nonlinear Control Systems II*. Springer-Verlag, 1999.

[120] V.I. Utkin. Variable structure systems with sliding modes. *IEEE Transactions on Automatic Control*, AC-22(2):212–222, 1977.

[121] J.Y. Hung, W.B. Gao, and J.C. Hung. Variable structure control: A survey. *IEEE Transactions on Industrial Electronics*, 40(1):2–22, 1993.

[122] X. Yu and O. Kaynak. Sliding-mode control with soft computing: A survey. *IEEE Transactions on Industrial Electronics*, 56(9):3275–3285, 2009.

[123] H.H. Choi. LMI-based sliding surface design for integral sliding model control of mismatched uncertain systems. *IEEE Transactions on Automatic Control*, 52(4):736–742, 2007.

[124] K.S. Kim, Y. Park, and S.H. Oh. Designing robust sliding hyperplanes for parametric uncertain systems: A Riccati approach. *Automatica*, 36(7):1041–1048, 2000.

[125] J.L. Chang. Dynamic output integral sliding-mode control with disturbance attenuation. *IEEE Transactions on Automatic Control*, 54(11):2653–2658, 2009.

[126] H.H. Choi. An explicit formula of linear sliding surfaces for a class of uncertain dynamic systems with mismatched uncertainties. *Automatica*, 34(8):1015–1020, 1998.

[127] P. Park, D.J. Choi, and S.G. Kong. Output feedback variable structure control for linear systems with uncertainties and disturbances. *Automatica*, 43(1):72–79, 2007.

[128] J. Xiang, W. Wei, and H. Su. An ILMI approach to robust static output feedback sliding mode control. *International Journal of Control*, 79(1):959–967, 2006.

[129] W.J. Cao and J.X. Xu. Nonlinear integral-type sliding surface for both matched and unmatched uncertain systems. *IEEE Transactions on Automatic Control*, 49(8):1355–1360, 2004.

[130] Y. Feng, X. Yu, and Z. Man. Non-singular terminal sliding mode control of rigid manipulators. *Automatica*, 38(12):2159–2167, 2002.

[131] Y.B. Shtessel, I.A. Shkolnikov, and A. Levant. Smooth second-order sliding modes: Missile guidance application. *Automatica*, 43(8):1470–1476, 2007.

[132] Y.B. Shtessel, I.A. Shkolnikov, and A. Levant. Guidance and control of missile interceptor using second-order sliding modes. *IEEE Transactions on Aerospace and Electronic Systems*, 45(1):110–124, 2009.

[133] J. Yang, S.H. Li, X.S. Chen, and Q. Li. Disturbance rejection of dead-time processes using disturbance observer and model predictive control. *Chemical Engineering Research and Design*, 89(2):125–135, 2011.

[134] A. Pomerleau, D. Hodouin, A. Desbiens, and E. Gagnon. A survey of grinding circuit control methods: from decentralized PID controllers to multivariable predictive controllers. *Powder Technology*, 102(2):103–115, 2000.

[135] A.J. Niemi, L. Tian, and R. Ylinen. Model predictive control for grinding systems. *Control Engineering Practice*, 5(2):271–278, 1997.

[136] M. Ramasamy, S.S. Narayanan, and C.D.P. Rao. Control of ball mill grinding circuit using model predictive control scheme. *Journal of Process Control*, 15(3):273–283, 2005.

[137] X.S. Chen, J.Y. Zhai, S.H. Li, and Q. Li. Application of model predictive control in ball mill grinding circuit. *Minerals Engineering*, 20(11):1099–1108, 2007.

[138] X.S. Chen, Q. Li, and S.M. Fei. Constrained model predictive control in ball mill grinding process. *Powder Technology*, 186(1):31–39, 2008.

[139] M. Duarte, F. Sepelveda, A. Castillo, A. Contreras, V. Lazcano, P. Gimenez, and L. Castelli. A comparative experimental study of five multivariable control strategies applied to a grinding plant. *Powder Technology*, 104(1):1–28, 1999.

[140] A.V.E. Conradie and C. Aldrich. Neurocontrol of a ball mill grinding circuit using evolutionary reinforcement learning. *Minerals Engineering*, 14(10):1277–1294, 2001.

[141] J.J. Govindhasamy, S.F. McLoone, G.W. Irwin, J.J. French, and R.P. Doyle. Neural modelling, control and optimisation of an industrial grinding process. *Control Engineering Practice*, 13(10):1243–1258, 2005.

[142] V.R. Radhakrishnan. Model based supervisory control of a ball mill grinding circuit. *Journal of Process control*, 9(3):195–211, 1999.

[143] X.S. Chen, S.H. Li, J.Y. Zhai, and Q. Li. Expert system based adaptive dynamic matrix control for ball mill grinding circuit. *Expert Systems with Applications*, 36(1):716–723, 2009.

[144] D. Hodouin, S.L. Jamsa-Jounela, M.T. Carvalho, and L. Bergh. State of the art and challenges in mineral processing control. *Control Engineering Practice*, 9(9):995–1005, 2001.

[145] C.R. Culter and D.L. Ramaker. Dynamic matrix control—A computer control algorithm. In *Proceedings of American Control Conference*, pages 281–286, 1980.

[146] H.W. Lee, K.C. Kim, and J. Lee. Review of Maglev train technology. *IEEE Transactions on Magnetics*, 42(7):1917–1925, 2006.

[147] K. Michail, A. Zolotas, and R. Goodall. Optimised sensor configurations for a Maglev suspension system. *IFAC Proceedings*, 17(1):169–184, 2008.

[148] R. Goodall. Dynamics and control requirements for EMS Maglev suspensions. In *Proceedings of International Conference on Maglev*, CD-ROM, 2004.

[149] R. Goodall. Generalised design models for EMS Maglev. In *Proceedings of International Conference on Maglev*, CD-ROM, 2008.

[150] J.S. Park, J.S. Kim, and J.K. Lee. Robust control of Maglev vehicles with multimagnets using separate control techniques. *KSME International Journal*, 15(9):1240–1247, 2001.

[151] P. Pillay and R. Krishnan. Control characteristics and speed controller design of a high performance PMSM. In *IEEE Industrial Application Society Annual Meeting*, pages 627–633, 1985.

[152] K.H. Kim and M.J. Youn. A nonlinear speed control for a PM synchronous motor using a simple disturbance estimation technique. *IEEE Transactions on Industrial Electronics*, 49(3):524–535, 2002.

[153] B. Grcar, P. Cafuta, M. Znidaric, and F. Gausch. Nonlinear control of synchronous servo drive. *IEEE Transactions on Control Systems Technology*, 4(2):177–184, 1996.

[154] G.J. Wang, C.T. Fong, and K.J. Chang. Neural-network-based self tuning PI controller for precise motion control of PMAC motors. *IEEE Transactions on Industrial Electronics*, 48(2):408–415, 2001.

[155] N. Golea, A. Golea, and M. Kadjoudy. Robust MARC adaptive control of PMSM driver under general parameters uncertainties. In *IEEE International Conference on Industrial Technology*, pages 1533–1537, 2006.

[156] S.H. Li and Z.G. Liu. Adaptive speed control for permanent magnet synchronous motor system with variations of load inertia. *IEEE Transactions on Industrial Electronics*, 56(8):3050–3059, 2009.

[157] Y.A.R.I. Mohamed and E.F. El-Saadany. A current control scheme with an adaptive internal model for torque ripple minimization and robust current regulation in PMSM drive systems. *IEEE Transactions on Energy Conversion*, 23(1):92–100, 2008.

[158] H.Z. Jin and J.M. Lee. An RMRAC current regulator for permanent-magnet synchronous motor based on statistical model interpretation. *IEEE Transactions on Industrial Electronics*, 56(1):169–177, 2009.

[159] T.-L. Hsien, Y.-Y. Sun, and Tsai M.-C. H_∞ control for a sensorless permanent-magnet synchronous drive. *IEE Proceedings—Electric Power Applications*, 144(3):173–181, 1997.

[160] R.J. Wai. Total sliding-mode controller for PM synchronous servo motor drive using recurrent fuzzy neural network. *IEEE Transactions on Industrial Electronics*, 48(5):926–944, 2001.

[161] S.H. Li, K. Zong, and H.X. Liu. A composite speed controller based on a second-order model of permanent magnet synchronous motor system. *Transactions of the Institute of Measurement and Control*, 33(5):522–541, 2011.

[162] B. Pioufile. Comparison of speed nonlinear control strategies for the synchronous servo-motor. *Transactions of the Institute of Measurement and Control*, 21(2):151–169, 1993.

[163] M. Vilathgamuwa, M. A. Rahman, K. Tseng, and M. N. Uddin. Nonlinear control of interior permanent magnet synchronous motor. *IEEE Transactions on Industry Applications*, 39(2):408–415, 2003.

[164] J. Zhou and Y. Wang. Adaptive backstepping speed controller design for a permanent magnet synchronous motor. *IEE Proceedings—Electric Power Applications*, 149(2):165–172, 2002.

[165] Y. Luo, Y.Q. Chen, H.S. Ahnc, and Y.G. Pi. Fractional order robust control for cogging effect compensation in PMSM position servo systems: stability analysis and experiments. *Control Engineering Practice*, 18(9):1022–1036, 2010.

[166] Y.S. Kung and M.H. Tsai. FPGA-based speed control IC for PMSM drive with adaptive fuzzy control. *IEEE Transactions on Power Electronics*, 22(6):2476–2486, 2007.

[167] S.H. Li, H.X. Liu, and S.H. Ding. A speed control for a PMSM using finite-time feedback control and disturbance Compensation. *Transactions of the Institute of Measurement and Control*, 32(2):170–187, 2010.

[168] K. Ohnishi. A new servo method in mechatronics. *Transaction of the Japanese Society of Electrical Engineers*, 107(D):83–86, 1987.

[169] W.H. Chen, D.J. Ballance, P.J. Gawthrop, and J.J. Gribble. Nonlinear disturbance observer for two-link robotic Manipulators. *IEEE Transactions on Industrial Electronics*, 47(4):932–938, 2000.

[170] M.T. White, M. Tomizuka, and C. Smith. Improved track following in magnetic disk drives using a disturbance observer. *IEEE/ASME Transanctions on Mechatronics*, 5(1):3–11, 2008.

[171] H.S Lee and M. Tomizuka. Robust motion controller design for high-accuracy positioning systems. *IEEE Transactions on Industrial Electronics*, 43(1):48–55, 1996.

[172] F.J. Lin, X. Cai, R.F. Fung, and Y.C. Wang. Robustfuzzy neural network controller with nonlinear disturbance observer for two-axis motion control system. *IET-Control Theory and Applications*, 2(2):151–167, 2008.

[173] K.H. Kim, I.C. Baik, G.W. Moon, and M.J. Youn. A current control for a permanent magnet synchronous motor with a simple disturbance estimation scheme. *IEEE Transactions on Control Systems and Technology*, 7(5):630–633, 1999.

[174] J.S. Ko and B.M. Han. Precision position control of PMSM using neural network disturbance observer on forced nominal Plant. In *IEEE International Conference on Mechatronics*, pages 316–320, 2006.

[175] J.B. Su, W.B. Qiu, H.Y. Ma, and P.Y. Woo. Calibration-free robotic eye-hand coordination based on an auto disturbance-rejection controller. *IEEE Transactions on Robotics*, 20(5):899–907, 2004.

[176] D. Wu, K. Chen, and X. Wang. Tracking control and active disturbance rejection with application to noncircular Machining. *International Journal of Machine Tools and Manufacture*, 47(15):2207–2217, 2007.

[177] Y.X. Su, B.Y. Duan, C.H. Zheng, G.D. Chen, and J.W. Mi. Disturbance-rejection high-precision motion control of a Stewart platform. *IEEE Transactions on Control Systems and Technology*, 12(3):364–374, 2004.

[178] B. Sun and Z. Gao. A DSP-based active disturbance rejection control design for a 1-kW H-bridge DC-DC power converter. *IEEE Transactions on Industrial Electronics*, 52(5):1271–1277, 2005.

[179] Y.X. Su, C.H. Zheng, and B.Y. Duan. Automatic disturbances rejection controller for precise motion control of permanent-magnet synchronous motors. *IEEE Transactions on Industrial Electronics*, 52(3):814–823, 2005.

[180] G. Feng, L.P. Huang, and D.Q Zhu. A new robust algorithm to improve the dynamic performance on the speed control of induction motor drive. *IEEE Transactions on Power Electronics*, 19(6):1614–1627, 2004.

[181] J.F. Pan, N.C. Cheung, and J.M. Yang. Auto-disturbance rejection controller for novel planar switched reluctance motor. *IEE Proceedings—Electric Power Applications*, 153(2):307–315, 2006.

[182] Y. Xia, P. Shi, G.P. Liu, and D. Rees. Active disturbance rejection control for uncertain multivariable systems with time-delay. *IET Control Theory Applications*, 1(1):75–81, 2007.

[183] S. Matsutani, Y. Sumiyoshi, M. Ishida, A. Imura, and M. Fujitsuna. Optimal control of PMSMs using model predictive control with integrator. In *ICCAS-SICE*, pages 4847–4852, 2009.

[184] F. Morel, X.F. Lin-Shi, J.M. Retif, B. Allard, and C. Buttay. A comparative study of predictive current control schemes for a permanent-magnet synchronous machine drive. *IEEE Transactions on Industrial Electronics*, 56(7):2715–2728, 2009.

[185] P.C. Krause. *Analysis of Electric Machinery (2nd ed.)*. McGraw-Hill, 1995.

[186] R. Miklosovic and Z.Q. Gao. A robust two-degree-of-freedom control design technique and its practical application. In *39th IAS Annual Meeting on Industry Application Conference*, pages 1495–1502, 2004.

[187] H.J. Kim and D.H. Shim. Output tracking control design of a helicopter model based on approximate linearization. In *Proceedings of the 37th IEEE Conference on Decision and Control*, pages 3635–3640, 1998.

[188] A. Krupadanam, A. Annaswamy, and R. Mangoubi. Multivariable adaptive control design with applications to autonomous helicopters. *Journal of Guidance Control and Dynamics*, 25(5):843–851, 2002.

[189] E. Johnson and S. Kannan. Adaptive trajectory control for autonomous helicopters. *Journal of Guidance Control and Dynamics*, 28(3):524–538, 2005.

[190] A. Bogdanov and E. Wan. State-dependent Riccati equation control for small autonomous helicopters. *Journal of Guidance Control and Dynamics*, 30(1):47–60, 2007.

[191] H. Kim, D. Shim, and S. Sastry. Nonlinear model predictive tracking control for rotorcraft-based unmanned aerial vehicles. In *Proceedings of American Control Conference*, pages 3576–3581, 2002.

[192] D. Shim, H. Kim, and S. Sastry. Decentralized nonlinear model predictive control of multiple flying robots. In *Proceedings of the 42nd IEEE Conference on Decision and Control*, pages 3621–3626, 2003.

[193] C.J. Liu, W.H. Chen, and J. Andrews. Tracking control of small-scale helicopters using explicit nonlinear MPC augmented with disturbance observers. *Control Engineering Practice*, 3(20):258–268, 2012.

[194] B. Mettler, M. B. Tischler, and T. Kanade. System identification modeling of a small-scale unmanned rotorcraft for flight control design. *Journal of the American Helicopter Society*, 47(1):50–63, 2002.

[195] V. Gavrilets, B. Mettler, and E. Feron. Dynamic model for a miniature aerobatic helicopter. In *AIAA Guidance Navigation and Control Conference*, Montreal, Canada, 2001.

[196] L. Marconi and R. Naldi. Robust full degree-of-freedom tracking control of a helicopter. *Automatica*, 43(11):1909–1920, 2007.

[197] I. A. Raptis, K. P. Valavanis, and W. A. Moreno. A novel nonlinear backstepping controller design for helicopters using the rotation matrix. *IEEE Transactions on Control Systems Technology*, 19(2):1–9, 2010.

[198] E. Panteley and A. Loria. On global uniform asymptotic stability of nonlinear time-varying systems in cascade. *Systems and Control Letters*, 33(2):131–138, 1998.

[199] S.Y. Lee, J.I. Lee, and I.J. Ha. Nonlinear autopilot for high maneuverability of bank-to-turn missiles. *IEEE Transactions on Aerospace and Electronic Systems*, 37(4):1236–1253, 2009.

[200] R. Whalley and M. Ebrahimi. The dynamic analysis of mechanical systems and structures with large parameter variations. *Systems and Control Letters*, 41(3):189–199, 1997.

[201] G.R. Duan, H.Q. Wang, and H.S. Zhang. Parameter design of smooth switching controller and application for bank-to-turn missiles. *Aerospace Control*, 23(2):41–46, 2005 (in Chinese).

[202] S. Komada, N. Machii, and T. Hori. Control of redundant manipulators considering order of disturbance observer. *IEEE Transaction on Industrial Electronics*, 47(2):413–4201, 2000.

[203] T.H. Liu, Y.C. Lee, and Y.H. Chung. Adaptive controller design for a linear motor control system. *IEEE Transaction on Aerospace Electronics Systems*, 40(2):601–616, 2004.

[204] S.N. Wu, X.Y. Sun, Z.W. Sun, and X.D. Wu. Sliding-mode control for staring-mode spacecraft using a disturbance observer. *Proceedings of the Institution of Mechanical Engineers Part G, Journal of Aerospace Engineering*, 224(G2):215–224, 2000.

[205] A.R. Mehrabian and J. Roshanian. Skid-to-turn missile autopilot design using scheduled eigenstructure assignment technique. *Proceedings of the Institution of Mechanical Engineers Part G, Journal of Aerospace Engineering*, 220(G3):225–239, 2006.

[206] D.O. Sigthorsson, P. Jankovsky, A. Serrani, S. Yurkovich, M.A. Bolender, and D.B. Doman. Robust linear output feedback control of an airbreathing hypersonic vehicle. *Journal of Guidance Control and Dynamics*, 31(4):1052–1066, 2008.

[207] B. Xu, F. Sun, C. Yang, D. Gao, and J. Ren. Adaptive discrete-time controller design with neural network for hypersonic flight vehicle via back-stepping. *International Journal of Control*, 84(9):1543–1552, 2011.

[208] B. Xu, D.X. Gao, and S.X. Wang. Adaptive neural control based on HGO for hypersonic flight vehicles. *Science China-Information Science*, 54(3):511–520, 2011.

[209] D.X. Gao and Z.Q. Sun. Fuzzy tracking control design for hypersonic vehicles via T-S model. *Science China-Information Science*, 54(3):521–528, 2011.

[210] B. Jiang, Z. Gao, P. Shi, and Y. Xu. Adaptive fault-tolerant tracking control of near-space vehicle using Takagi-Sugeno fuzzy models. *IEEE Transactions on Fuzzy Systems*, 18(5):1000–1007, 2010.

[211] Y.N. Hu, Y. Yuan, H.B. Min, and F.C. Sun. Multi-objective robust control based on fuzzy singularly perturbed models for hypersonic vehicles. *Science China-Information Science*, 54(3):563–576, 2011.

[212] Z. Gao, B. Jiang, P. Shi, and Y. Xu. Fault-tolerant control for a near space vehicle with a stuck actuator fault based on Takagi-Sugeno fuzzy model. *Proceedings of the Institution of Mechanical Engineers Part I-Journal of Systems and Control Engineering*, 224(I5):587–597, 2010.

Index